CAMBRIDGE COUNTY GEOGRAPHIES

General Editor: F. H. H. GUILLEMARD, M.A., M.D.

T0352321

WARWICKSHIRE

Cambridge County Geographies

WARWICKSHIRE

by

J. HARVEY BLOOM, M.A.

With Maps, Diagrams and Illustrations

Cambridge :
at the University Press
1916

CAMBRIDGE UNIVERSITY PRESS
Cambridge, New York, Melbourne, Madrid, Cape Town,
Singapore, São Paulo, Delhi, Mexico City

Cambridge University Press
The Edinburgh Building, Cambridge CB2 8RU, UK

Published in the United States of America by Cambridge University Press, New York

www.cambridge.org
Information on this title: www.cambridge.org/9781107611443

First published 1916
First paperback edition 2013

A catalogue record for this publication is available from the British Library

ISBN 978-1-107-61144-3 Paperback

PREFACE

A FEW short sentences will suffice to express the thanks of the author to the many kind friends who have assisted in the preparation of this little volume. Mr E. S. Scott, late Science Master at Stratford Grammar School, whose early death is deeply regretted, was responsible for the bulk of the photographs, which he took expressly for this work. Photographs were also contributed by Mr Sale of Atherstone and Mr F. A. Newdegate, M.P., and various industrial firms. Mr F. S. Potter of Halford advised the author throughout and kindly read the proofs; Mr Edwin Smith of Great Alne also gave kind advice and made valuable suggestions. Thanks are also due to the General Editor of this series and to Mr S. C. Roberts, as also to various city and county officials and to the director of the Midlands Institute Meteorological Station at Edgbaston.

J. H. BLOOM.

October 1915.

CONTENTS

ILLUSTRATIONS

MAPS

For the photographs reproduced on pp. 7, 13, 14, 26, 27, 75, 76, 98, 132, 136, 138, and 140 thanks are due to the late Mr E. S. Scott; for those on pp. 96 and 134 to Mr G. E. Over, Rugby; for that on p. 53 to Mr F. A. Newdegate, M.P.; for that on p. 48 to Mr H. Sale, Atherstone Hall; for that on p. 74 to Mr Edwin Smith; for those on pp. 52, 55, and 56 to the Kingsbury Collieries, Ltd, the B.S.A. Co., and the Humber Co. respectively.

The photographs reproduced on pp. 17, 19, 78, 80, 81, 83, 84, 89, 93, 102, 113, 115, 117, 128, and 139 were supplied by Messrs F. Frith and Co., Reigate; those on pp. 67, 88, 92 and 95 by Mr L. C. Keighley-Peach; those on pp. 69 and 103 by Messrs J. Valentine and Sons, Dundee; those on pp. 44 and 47 by Mr W. A. Smith; those on pp. 31 and 33 by Mr T. R. Hodges, Fleet; those on pp. 120 and 121 by Messrs Emery Walker, Ltd; that on p. 91 by Messrs W. Cawthorne and Son, Nuneaton; that on p. 11 by Mr W. R. Sansbury, Banbury; that on p. 4 by Mr W. Stephenson; that on p. 72 by Mr F. C. Welstood.

1. County and Shire.

The study of the geography of a county is not completed by a knowledge of the general character of its rocks and soils, nor even of its chief physical features. Man, during his long struggle with the forces of nature, has not only altered the surface by clearing forests, draining marshes, etc., but has turned large tracts of heath and moorland into rich cornlands and upland pastures. He is still actively engaged in mining rocks below the surface and demolishing hills above it, in order to obtain the motive power for the many factories which he has raised where forest trees once grew, or to procure material for miles of carefully constructed main roads and railway banks, over which to carry produce to distant parts. The whole country side has been altered by man. Great cities have sprung into being where there was once waste land. It is therefore necessary to learn something of the history of a county before we can understand its geography.

English counties often present great irregularity of outline, and Warwickshire is no exception. This irregularity is due to many causes, partly natural, partly artificial, but principally to the fact that our counties were not formed simultaneously, nor for similar reasons. Had they been thought out and laid down by rule, it is probable that natural boundaries would have been more uniformly made use of; or, as in the case of the United States, that the counties would have been delimited in rectangles by the surveyor's rod. Even when their bounds had once been decided, they were not free from alterations; indeed such alterations are still taking place whenever new conditions render them necessary. Certain English counties, such as Essex, Kent, or Norfolk, are complete in themselves, the relics of ancient Saxon kingdoms; but others, such as Warwickshire, are but a part of what was originally a far larger tract of country. In the case of our own county this tract was known as the kingdom of Mercia, and owed its origin to the conquests of a leader named Cridda, who became its first king.

The word county (*comté*, the land of a *Comte* or count) is French, and was brought over by the Normans. It was a useful word to express a tract of country treated as a unit for the purposes of administration. This word county to a certain extent displaced the older native word shire, still pronounced by the peasantry shear, a word (from the Anglo-Saxon *sciran*) which meant a portion of land shorn or cut off from a larger whole, as Warwickshire was cut off from Mercia. The word county is used in a geographical sense in such expressions as County Council, County Court, County town, while shire is retained when we have occasion to speak of the Shire Hall or the Sheriff (Shire Reeve). It is therefore quite

correct to speak of the County of Warwick, or of Warwickshire, but not correct to say the County of Warwickshire.

This idea of cutting up England into counties is said
to have been planned by Alfred the Great as a military
protection against the Danes. There was to be a county
town near the centre of each shire, with a garrison
always in readiness to resist a sudden raid, furnished in
relays by every town and village in the county. This
was called *Burh-bot*. The town chosen often became
the capital of the county, and gave its name to it, as in
the case of Warwick, Worcester, Gloucester, or Northampton.

The name Warwick was in existence as far back as
A.D. 701; but there is no mention of the county before the
year 1016, so that it is not at all likely that our county
was formed by Alfred. It is more probably the work of
Edward the Elder, and resulted from his victory over the
Danes. The meaning of Warwick (Waerinc Wicum)
is village of the Waerings.

2. General Characteristics.

Warwickshire lies a little west of the centre of
England, and is therefore entirely an inland district. It
possesses at the present day no navigable river, though at
one time, not very remote, barges from the Severn were
able to unload at Stratford-on-Avon.

The county is enclosed by low hills along its western

and south-western borders, and a series of raised uplands of varied character cover the greater part of its northern area. To much of this the term plateau has been applied, but it is in no sense a tableland. On the east the valley of the Avon widens into a broad plain, which is nearly a

Typical Warwickshire Scenery
(*Arlscote—the north-eastern corner of the Edge Hills*)

uniform level, watered by numerous small streams; and on the extreme south-eastern border the Edge Hills divide the county from Oxfordshire.

The greater part of the northern half is richly wooded, and it possesses beautiful parks and wide stretches of pasture land. The wooded character of this portion led the older writers to describe it under the name of the Forest of Arden, although there was no legal or technical forest in the county except such portions of the Forest of Feckenham as strayed over from Worcestershire to the banks of the Arrow. It was also known as the Weldon, in contradistinction to the portion south of the Avon called the Feldon, which was less well wooded. In the far north Sutton Park or Chace still retains the wild character which its latter name suggests.

The forest-like character of the landscape is found wherever the Keuper marl forms the subsoil, and in these districts the oak and ash, our noblest native trees, flourish. In other districts they are replaced by the introduced elm, a beautiful tree in early summer, but assuming later a sombre hue which gives a somewhat gloomy look to the country-side.

In the past the county formed part of a second line of defence, designed as a protection from Welsh marauders ; the first line depending upon the moral and religious influence of the great monasteries of Gloucestershire and Worcestershire more than upon force of arms. But behind the chain of religious houses ran another chain of strong military posts, of which our Warwickshire Castles, Tamworth, Kenilworth, and Warwick, are the most important, probably all of them dating from the foundation of the Mercian Kingdom, and thrown up as outposts of Mercia in the struggle with the Danes.

The central position of the county made it the Belgium of medieval England. Troops passed and re-passed through it from the wars of Stephen and Maud to the northern march of the Duke of Cumberland, but it was never called upon to resist any invasion from Wales. In its earliest days a chain of camps ran along its hill-tops, or defended important posts by the principal fords ; and possibly on the north the primeval forest offered an even stronger defence until the Romans penetrated it, and cut roadways through it.

The greater portion of its surface is now farmed as pasture, and agricultural pursuits occupy the time and energy of its people, save only in the north of the county, where the presence of valuable coal-seams and other minerals have caused the vast industrial development of the cities of Birmingham and Coventry and the towns of Atherstone and Nuneaton, while Rugby in the east forms a detached centre of commercial activity.

Much of the county is marked by the red soil of the Keuper, and this, when newly ploughed, glows in the spring sunshine as vividly as vermilion and lends a rich warmth to the landscape. The homes of the people are also warm in colour. The older houses are constructed largely of timber, wherever that material was plentiful, but in the south they gradually assume the Cotswold character, being built of yellow lias and oolite which becomes richly toned with lichen, while the thatched roofs of the north are replaced by stone tiles.

The industrial portion is becoming more and more a land of villas wherein the wealthier traders of the great cities dwell. These are usually brick-built with roofs of machine-made tiles and are not as yet objects of beauty, though they evince no inconsiderable sign of prosperity.

Cottage at Upton-in-Haselor
(*a half-timbered house*)

The Falcon Inn, Bidford
(*a stone-built house*)

3. Size. Shape. Boundaries.

The county lies between 51° 57′ 30″ and 52° 42′ north latitude and between 1° 7′ 30″ and 1° 56′ 40″ west longitude. It has a total length of some 52 miles from Newton Regis in the north to Long Compton in the south, and from the outskirts of Redditch on the west to Hillmorton on the east it is about 33 miles in width. The very irregular boundaries cover a length of about 200 miles, exclusive of a small separated portion of the county lying on the south-west which is eight miles in length by three in width. The Geographical County has an area of 902 square miles, or 577,462 acres. It is therefore larger than Cambridgeshire, but smaller than Dorsetshire, the counties nearest to it in size.

In shape Warwickshire somewhat resembles a pear with its broad end to the north. Its outline, however, is broken on its western side by two deep indentations formed by portions of Worcestershire; one of these runs in a north-easterly direction towards Coleshill; the other, which is much narrower, in a south-easterly direction towards Henley-in-Arden. Two portions are outlying parts of Gloucestershire parishes—Bickmarsh, a hamlet of Welford-upon-Avon, and Milcote, a hamlet of Weston-upon-Avon. The former is separated from its mother parish by some miles. In 1842 two detached parishes, also belonging to Gloucestershire, were added to Warwickshire, namely Little Compton, and the island of Sutton-under-Brailes, which anciently belonged to Westminster

Abbey and was reckoned with its other properties outside the county. In the same year Tutnal and Cobley, Warwickshire hamlets, were transferred to Worcestershire, and other insignificant changes have since taken place.

The county is bounded on the north-east by Leicestershire, on the east by Northamptonshire, on the southeast by Oxfordshire, on the south-west by Gloucestershire, on the west by Worcestershire, and on the northwest by the county of Stafford. The boundary line follows at times natural features, at others the course of an early road, but more often its course is very irregular, and to appearance without definite plan.

Starting from No Man's Heath at the north it follows an irregular course to the river Anker, with which it coincides from near Grendon to Mancetter. Thence to its most easterly point, save for a little space near Nuneaton, the Watling Street serves to bound it for a distance of 21 miles. It then follows one of the tributaries of the Leam to the neighbourhood of Willoughby, and, crossing the Oxford Canal a little beyond, rises on to the high ground in the neighbourhood of Shuckburgh. It retains this elevated position to the extreme southern end of the county, except at one or two places, averaging a height of between 400 and 700 feet, and pursuing a general S.S.W. direction. In its course it passes the Three Shire Stones near Wormleighton, includes the *massif* of the Edge Hills and the beautiful seat of Compton Wynyates, crosses at Pitch Hill the road from Banbury to Shipston-upon-Stour, and takes a straight and almost southerly course for some five miles towards Great

Rollright village. This, however, together with the remarkable Neolithic circle known as Rollright Stones, it just avoids, leaving them in Oxfordshire, and a mile or two beyond reaches its most southerly point, which is on the road connecting Banbury with Stow-on-the-Wold, about five miles from the latter town. The boundary now turns northwards, and we approach Moreton-in-the-Marsh, at about a mile to the east of which is the Four Shire Stone, where Oxfordshire, Gloucestershire, Warwickshire, and part of Worcestershire meet. From this landmark the boundary is again irregular to Nethercot Brook, which it follows until it joins its waters to the Stour at Burmington. The line then lies along the Stour to Ettington, when it turns inland to give place to Worcestershire. This county in turn gives way for a brief space to a Gloucestershire parish (Preston) when Warwickshire for a time takes possession of both banks of the river, to be ousted by the Gloucestershire parish of Clifford, which later gives way to Stratford-on-Avon.

When Worcestershire passes the Stour, Warwickshire crosses over to form a considerable island, embracing the parishes of Ilmington, Whitchurch, and Stretton-on-the-Fosse.

The right bank of the Avon mainly forms the boundary from Stratford to Salford Priors: thence to Weethley Cross no natural feature delimits it, but at that point it touches the Ridgeway and follows the road nearly to Redditch, after which it is again irregular until it meets the river Cole at Coleshill. It then bends round

Birmingham to the Icknield Way at Sutton Park, and
thence turns eastwards to the river Tame at Dosthill

The Four Shire Stone

which it follows to Tamworth.   Thence taking a north-
easterly direction it reaches the northernmost point of the
county from the neighbourhood of which we started.

The detached portion of the county already referred to was part of the vast estates of the Count of Meulan, and was thus assessed with the rest of his possessions. Other portions belonged to Osbern son of Richard, owner of Richard's Castle, and a friend of St Edward the Confessor, and to Robert of Stratford, who also held estates in other parts of Warwickshire, so that it was convenient for geld (taxation) purposes that these properties should be reckoned as part of the county.

4. Surface and General Features.

A map in relief of Warwickshire shows it to be an undulating plain with no lofty heights or jagged mountains. There are occasional steep and sudden slopes, however, as there are remarkable bald hills. The one striking feature is the great escarpment of the Edge Hills, which looks almost knife-like from a distance; but there is nothing like the Malvern Hills or the Lickey. The scenery is typically English, pleasant woodlands, fertile valleys, an occasional stretch of cornland, and meadows everywhere.

A patch or two of heath occurs at Wolford in the extreme south, another yet smaller at Snitterfield, and other and larger stretches in Sutton Park, but there is nothing to match the colour of the Wyre Forest, and even an expanse of golden gorse is hard to find. There are no bogs or marshes of any size. The county has been too well cultivated to allow of their existence.

Nor are there any well-defined special areas, such as the Forest of Dean, or the Clent Hills, although the general character of the district north of the Avon has already been noticed. The county whether north or south is largely a land of trees, but it must not be

Burton Dassett Hills

forgotten that most of these are of modern planting. The old tapestry map of the county, now in the Bodleian Library, woven at Barcheston about 1600, shows only two deer parks, of which Wedgenock, now nearly destroyed, was perhaps the oldest and most considerable. Fulbrook, the other, has long since gone.

Very fine views can be obtained from Burton Dassett windmill, from Upton Tower on Edge Hill, or from Brailes Hill, but to realise the beauty of the county at its best, there are few better spots than the Alne Hills, which enable the observer to see for a distance of from

Burton Dassett Beacon and Windmill

30 to 40 miles in all directions, and give a bird's-eye view of the valleys through which the Blythe, the Alne, the Arrow, the Avon, and the Stour flow, while the scene is backed by the pointed masses of the Malverns, the rounded hump of Bredon, and the ranges of the Lickey, the Clent, and the Cotswolds. From Alne Hills the

county looks like a wooded basin set in a rim of hills, saving only towards Rugby.   A very similar effect can be witnessed at Hampton-in-Arden, but upon a far smaller scale.

On approaching Birmingham the scene alters ; lofty factory shafts rise among the trees, and the houses become more and more densely suburban.   In the last few years the houses have eaten up the country districts at an alarming rate, not only about Birmingham but at Nuneaton and Coventry also.   As the train takes us to the east the black smoke of collieries and the desolate-looking rubbish-heaps mar the landscape on all sides until we reach Nuneaton, where they stand thickest.   Yet even here Warwickshire asserts herself, and the beautiful park of Arbury with its oaks and pines and large pools is there to speak for it.

In the old days of danger and alarum the county was easily roused ; the beacons on Burton Dassett Hill, Bickenhill, and at Monks Kirby amply sufficed to warn inhabitants of the coming foe.

5.   Watersheds and River Basins.

The river-system of Warwickshire is a simple one. The greater part of the county is drained by the Avon, which is itself a tributary of the Severn.   The other chief portion drains by the Tame into the Trent, while a small and inconsiderable area lies in the basin of the Thames. It is thus remarkable that rain falling in the county may ultimately find its way to such widely separated waters as the Bristol Channel, the North Sea, and the mouth of the English Channel.

To deal with this latter portion first we find it lies in the extreme south-east, where certain small brooks flow from the Edge Hills to feed the Cherwell, a river which in its turn empties itself into the Thames at Oxford. The slope drained is a very gradual one, ending in a wide extending plain.

That part of the county within the basin of the Trent lies to the north. It has no lofty heights, but is drained by several rivers, of which the Tame is the largest. This stream rises at Bloxwich in Staffordshire, and enters the county at Witton, on the northern out-skirts of Birmingham, where it is little more than a brook. Various streams from Sutton Park increase its volume, and at Ham Hall it receives the waters of the Blythe and the little stream called the Bourne. These drain a considerable area about Whitacre. At Tamworth the Anker joins it and it then flows on under Ladybridge to mingle with the Trent near Croxall in Derbyshire.

Of its tributaries, the Blythe rises at the height of 585 feet among the western hills, passing through Earlswood reservoir to Hampton-in-Arden and Coleshill, and receiving several small streams on its course. The Cole, its other branch, also rises among the hills, and forms for a short distance the boundary of the county, turning into it near Sheldon Hall, and so past Coleshill to join the Blythe at Blythe Hall. Both these streams run through level meadows and are liable to flood.

Charlecote Park

The longest and most important river is the Avon, a stream which divides Warwickshire into two more or less equal halves, and is in many places conspicuous for its beauty, flowing, as it does, through overhanging woods and pleasant meadows. The main stream rises at Naseby in Northamptonshire, enters the county at Clifton, and flows by way of Rugby to Stoneleigh Abbey, where it is joined by the Sow. Emscote, between Warwick and Leamington, is then reached, and here the Leam pours its waters into it, after which it passes through Warwick, where it is spanned by a picturesque ruined bridge. At Hampton Lucy Thelesford Brook adds to its waters, and at Charlecote Park the stream called the Dene. The Avon then flows through the park to Hatton Rock, its most beautiful portion, and under the great bridge of fourteen arches built at Stratford by Sir Hugh Clopton, to its confluence with the Stour a mile beyond. At Binton the "Three Bridges" cross it, and at Bidford the old fifteenth century bridge built by the monks of Bordesley. It is then joined by the Arrow coming from the north, and shortly after leaves the county near Salford on its way to join the Severn at Tewkesbury. Throughout its length its banks are lined with many beautiful plants. The flowering rush and water arrow-head abound, and masses of yellow water lilies. Like most Warwickshire rivers it is prone to flood, and at such times the levels of its valley become a mimic sea.

It is now time to speak of its tributaries. Of these the Leam rises on the northern slope of Marston Hill, flowing along the county boundary until it enters it at

Wolfhamcote. It is joined near Grandborough by the Rains Brook, and shortly after leaving Marton by the Itchen. It is this stream that forms the chief attraction of the ornamental gardens at Leamington.

The Sow rises in the north of the county about a mile north of Monks Kirby, and drains much of the mining

The Avon at Stratford

district, following an irregular course until it receives the Sherburne at Baginton, and joining the Avon in the beautiful park of Stoneleigh Abbey.

The Alne, another feeder, is formed by the confluence of two streams which rise on the border, about three miles N.E. of Redditch. It flows past Tanworth to

2—2

Wootton Wawen, where it is met by a branch from
Wroxall Abbey, it then turns to meet the Arrow at
Alcester. The last-named stream rises in a valley near
Alvechurch in Worcestershire, crossing into Warwickshire near Redditch. It then passes Studley, Coughton
and Alcester, and finally joins the Avon at Salford. Much
of its course is prettily wooded.

The most important tributary of the Avon is the
Stour, which drains all the south of the county. It
comes to us from Stour Well, and enters Warwickshire
at Traitor's Ford, winding among the hills past Stourton,
through Shipstone to Honington Park, and thence under
Halford Bridge to Ettington and Alscot Parks, eventually
winding through a wide level valley to the Avon at
Milcote. It is very apt to flood, and its waters rise with
great rapidity and cover a wide extent of level meadow-
land. This is partly due to the height of its watershed,
which includes Ebrington Hill (855 feet), Bright Hill
(737 feet), and Brailes Hill (700 feet).

There are no Warwickshire lakes, but important pools
exist at Packington, Olton, Merecot, Coombe Abbey,
and Warmington. The pools at Holt, Chesterton, and
Bishop's Itchington, close by, are brackish, and marine
plants, to be referred to later, grow around them.

6. Geology.

By Geology we mean the study of the rocks, and
we must at the outset explain that the term *rock* is used
by the geologist without any reference to the hardness

	Names of Systems	Subdivisions	Characters of Rock.
TERTIARY	**Recent Pleistocene**	Metal Age Deposits Neolithic ,, Palaeolithic ,, Glacial ,,	Superficial Deposits
	Pliocene	Cromer Series Weybourne Crag Chillesford and Norwich Crags Red and Walton Crags Coralline Crag	Sands chiefly
	Miocene	Absent from Britain	
	Eocene	Fluviomarine Beds of Hampshire Bagshot Beds London Clay Oldhaven Beds, Woolwich and Reading Thanet Sands [Groups	Clays and Sands chiefly
SECONDARY	**Cretaceous**	Chalk Upper Greensand and Gault Lower Greensand Weald Clay Hastings Sands	Chalk at top Sandstones, Mud and Clays below
	Jurassic	Purbeck Beds Portland Beds Kimmeridge Clay Corallian Beds Oxford Clay and Kellaways Rock Cornbrash Forest Marble Great Oolite with Stonesfield Slate Inferior Oolite Lias—Upper, Middle, and Lower	Shales, Sandstones and Oolitic Limestones
	Triassic	Rhaetic Keuper Marls Keuper Sandstone Upper Bunter Sandstone Bunter Pebble Beds Lower Bunter Sandstone	Red Sandstones and Marls, Gypsum and Salt
PRIMARY	**Permian**	Magnesian Limestone and Sandstone Marl Slate Lower Permian Sandstone	Red Sandstones and Magnesian Limestone
	Carboniferous	Coal Measures Millstone Grit Mountain Limestone Basal Carboniferous Rocks	Sandstones, Shales and Coals at top Sandstones in middle Limestone and Shales below
	Devonian	Upper Mid } Devonian and Old Red Sand- Lower stone	Red Sandstones, Shales, Slates and Limestones
	Silurian	Ludlow Beds Wenlock Beds Llandovery Beds	Sandstones, Shales and Thin Limestones
	Ordovician	Caradoc Beds Llandeilo Beds Arenig Beds	Shales, Slates, Sandstones and Thin Limestones
	Cambrian	Tremadoc Slates Lingula Flags Menevian Beds Harlech Grits and Llanberis Slates	Slates and Sandstones
	Pre-Cambrian	No definite classification yet made	Sandstones, Slates and Volcanic Rocks

or compactness of the material to which the name is applied; thus he speaks of loose sand as a rock equally with a hard substance like granite.

Rocks are of two kinds, (1) those laid down mostly under water, (2) those due to the action of fire.

The first kind may be compared to sheets of paper laid one over the other. These sheets are called *beds*, and such beds are usually formed of sand (often containing pebbles), mud or clay, and limestone, or mixtures of these materials. They are laid down as flat or nearly flat sheets, but may afterwards be tilted as the result of movement of the earth's crust, just as we may tilt sheets of paper, folding them into arches and troughs, by pressing them at either end. Again, we may find the tops of the folds so produced worn away as the result of the wearing action of rivers, glaciers, and sea-waves upon them, as we might cut off the tops of the folds of the paper with a pair of shears. This has happened with the ancient beds forming parts of the earth's crust, and we therefore often find them tilted, with the upper parts removed.

The other kinds of rocks are known as igneous rocks, and have been melted under the action of heat and become solid on cooling. When in the molten state they have been poured out at the surface as the lava of volcanoes, or have been forced into other rocks and cooled in the cracks and other places of weakness. Much material is also thrown out of volcanoes as volcanic ash and dust, and is piled up on the sides of the volcano. Such ashy material may be arranged in beds, so that it partakes to some extent of the qualities of the two great rock groups.

The production of beds is of great importance to
geologists, for by means of these beds we can classify the
rocks according to age.   If we take two sheets of paper,
and lay one on the top of the other on a table, the upper
one has been laid down after the other.   Similarly with
two beds, the upper is also the newer, and the newer will
remain on the top after earth-movements, save in very
exceptional cases which need not be regarded by us here,
and for general purposes we may regard any bed or set of
beds resting on any other in our own country as being
the newer bed or set.

The movements which affect beds may occur at
different times.   One set of beds may be laid down flat,
then thrown into folds by movement, the tops of the
beds worn off, and another set of beds laid down upon the
worn surface of the older beds, the edges of which will
abut against the oldest of the new set of flatly deposited
beds, which latter may in turn undergo disturbance and
renewal of their upper portions.

Again, after the formation of the beds many changes
may occur in them.   They may become hardened, pebble-
beds being changed into conglomerates, sands into sandstones, muds and clays into mudstones and shales, soft
deposits of lime into limestone, and loose volcanic ashes
into exceedingly hard rocks.   They may also become
cracked, and the cracks are often very regular, running in
two directions at right angles one to the other.   Such
cracks are known as *joints*, and the joints are very important
in affecting the physical geography of a district.   Then,
as the result of great pressure applied sideways, the rocks

may be so changed that they can be split into thin slabs, which usually, though not necessarily, split along planes standing at high angles to the horizontal. Rocks affected in this way are known as *slates*.

If we could flatten out all the beds of England, and arrange them one over the other and bore a shaft through them, we should see them on the sides of the shaft, the newest appearing at the top and the oldest at the bottom, much as in the table on p. 21. Such a shaft would have a depth of between 10,000 and 20,000 feet. The strata beds are divided into three great groups called Primary or Palaeozoic, Secondary or Mesozoic, and Tertiary or Cainozoic, and the lowest Primary rocks are the oldest rocks of Britain, and form as it were the foundation stones on which the other rocks rest. These are usually termed the Pre-Cambrian rocks. The three great groups are divided into minor divisions known as systems. The names of these systems are arranged in order in the table. On the right hand side, the general characters of the rocks of each system are stated.

With these preliminary remarks we may now proceed to a brief account of the geology of the county.

In a geological sense the greater part of the county is comparatively recent, but ancient rocks are not wanting, since they are represented near Nuneaton, and are rich in minerals having a considerable commercial value. The oldest rocks of the world, as we have seen, are called Pre-Cambrian or Archaean. These come to the surface in a narrow strip, scarcely a quarter of a mile wide and only two miles long, which starts from the Midland railway station at Nuneaton and runs in a north-westerly direction. The rocks were perhaps formed by a group of volcanoes, which at some extremely ancient date were active on islands of an archipelago then existent where Great Britain now is. Very gradually these volcanoes wore themselves out and eventually sank. A Cambrian sea then deposited its sand and mud, and the rocks so produced are found close by, at Hartshill and Stocking-ford, and may be traced for a few miles further by the picturesque outlines of the hills. These hard rocks have an economic value, being much used for road metal, while in them the earliest local fossils may be found, Trilobites and Brachiopods being often abundant.

In Warwickshire there are no beds to fill the long space of time between the end of the Cambrian period and the Carboniferous. During this period the former rocks slowly sank, leaving a vast morass but little raised above sea level. It was on this fen land that the coal-measures were laid down. The district is not large, some fifteen miles by four, reaching from Bedworth to Tam-worth. This does not mark its original extent, which was no doubt very great, covering South Staffordshire also. The latest Primary rocks—the Permian—are found extending from Baxterley in the north to Kenilworth in the south, and much building stone is quarried from them at Coventry, Kenilworth, and Warwick. From Kenil-worth came the Red Sandstone so conspicuous in the churches which belonged to that Priory. In one of these quarries the only known skeleton of *Dasyceps Bucklandi*, a fossil reptile, was found. At the end of the Permian period the swamps in which grew the plants of the Coal Age had disappeared, while the land had been considerably raised, to be exposed for a long period to the action of the atmosphere.

In the course of time the sandstones and marls of the

Lias Quarry
(*near Ettington*)

Trias were gradually deposited—the Bunter pebble-beds in desert lakes, the Keuper marl in an inland sea. The sandstones pass by slow degrees into the marls, which are used for brick-making. Near Alcester veins of gypsum occur in them, but the workings have lately ceased to be remunerative. In these rocks tracks of five-toed and three-toed reptiles occur. They are followed in point of time by the thin beds known as the Rhaetic, sometimes confused with the Lower Lias. Their fossils are very numerous, and some of the stone, the "Guinea Bed," rings when struck by the hammer like the coin after

Lias Cutting
(*near Ettington*)

which it is named. These Rhaetic rocks occur at Binton, Wootton, and Temple Grafton. They were probably deposited when these islands were sinking, so that the barriers which had kept out the ocean could no longer perform the work.

Coming now to the rocks of the Jurassic system, we

find that the Lias series succeeds the Rhaetic with no sign of any violent change. Its beds are usually remarkably even and consist of blue or white limestones and shales, and are generally rich in fossils. They have been much quarried at Wilmcote for cement, and also for paving-stones at Binton. There are also important lime-works at Harbury and Rugby. The Middle Lias enters the county on the south-east and forms the remarkable escarpment of the Edge Hills, which rises to the height of 716 feet above sea level : where it occurs the surface soil is rich and brown, and produces excellent wheat. These Lias rocks were laid down in shallow seas in which gigantic reptiles abounded, and fine skeletons of Ichthyosaurus have been found both at Honington and Wilmcote. The shallow seas gave way gradually to the deeper seas of the Upper Lias, which near Ilmington reaches a thickness of nearly 120 feet. At the close of the period the climate was warm and the seas had again become more shallow. The Oolites which follow are poorly represented. They occur here and there in patches near Ilmington and from Little Compton to Compton Wynyates, and are usually rich in fossils.

The geological history of the county is then for a long time a blank. The sea bottom slowly sank, and thousands of feet of strata must have been deposited over it, raised above water, and then slowly worn away by the action of the atmosphere and river currents, until the surface took more or less the form with which we are now familiar. Over this surface ice-sheets and glaciers deposited the sands, gravels, and clays which form the river gravels of the Pleistocene period. Three great glaciers met in the Midlands, one coming from the Arenig mountains in North Wales, another from the south of Scotland, and a third from the North Sea. Very gradually the climate grew warmer, and the melting ice broke up and deposited much of the older Drift. The present river system then came into being and began to wear down the valleys, and deposit along them the river terraces, in which man at length appears.

7. Natural History.

Long ages ago the land connections of what we now know as the British Isles were very different from those of to-day. Our land was then no group of islands, but formed a north-western projecting portion of the Continent. The greater part of its surface lay under a thick coating of ice, so that few if any of the pre-existing animals and plants can have survived. When the Glacial Period passed, there must have been a gradual re-peopling from the south and east by the various species. But not all of those then existing on the Continent had time to establish themselves in our land before the latter became isolated by the formation of the North Sea and the Channel. Hence we possess fewer species than the Continent; and Ireland—which was earlier cut off—still fewer.

Warwickshire being a central county, destitute of seaboard, is necessarily not so rich in species as many others more favourably placed. But, on the other hand, it lies midway between the Wash and the Bristol Channel, on an old line of migration, and hence attracts birds which would not otherwise be seen. The valley of the Avon, too, is specially favourable ground, rich in wild flowers and insects, which in their turn attract a varied fauna. The county, it is true, does not afford the striking diversity of feature that we find, for example, in Devonshire or in Cambridgeshire. It has no fens and no great moors like Yorkshire, yet numbers of water birds wander up the river valleys, as the decoys at Warwick and Arbury testify; and though the greatest elevation the county can boast is only 855 feet (Ebrington Hill), the grouse is a Warwickshire bird.

Of the 12 bats found in England eight have been recorded from our county, among them the lesser horseshoe bat (*Rhinolophus hipposiderus*), somewhat of a rarity. The three shrews—the common shrew, pygmy shrew, and water shrew—are all found, though it is said that the two latter are confined to the valley of the Avon. Of rodents, besides the generally-distributed hare, rabbit, squirrel, and brown rat (the black rat also occurs, though rarely), the common and long-tailed field mouse are abundant, but the little red harvest mouse is believed not to exist in the north of the county. The water, field, and bank voles are common, so are the mole and hedgehog.

Passing to the carnivores, the polecat, once fairly numerous, is now extinct. The weasel and stoat, like the fox, are abundant. The latter is hunted by two packs of hounds, the North and South Warwickshire. The badger is only rarely to be found, chiefly on the slopes of the Cotswolds, at Ilmington and Crimscote, and also at Burton Dassett. The otter, on the other hand, seems to be increasing and is fairly numerous in the Avon and the Stour.

By many persons it is thought that the swallow is much less abundant in England than it used to be, but whether the decrease is actual, or only apparent, owing to shifting of habitat, it is not easy to say. But there seems no doubt that while in former times these birds used to haunt the reeds on the Avon in numbers so great that the hobbies used to follow them to prey upon them, this is no longer the case. Warwickshire may be regarded in some ways as a border country, affording examples both of northern and of southern types. Thus the dipper is occasionally seen in the streams and the twite also occurs, though rarely. On the other hand the nightingale is found over the greater part of the county, especially in the low-lying tracts. Sutton Coldfield is still more or less of a bird sanctuary, and here have been shot the bittern and little bittern, and the lesser egret. Quail are now un-common, and the corncrake is said to be decreasing, but the hawfinch is getting commoner, and, of course, the starling, a bird which has increased its numbers and its range marvel-lously within the last twenty years or so. The ring-ousel occurs as a bird of double passage. There is still a heronry in the grounds of Warwick Castle, and a small one at Ragley, and

Slow-worm

the great crested grebe nests in more than one locality.

In suitable places in the valley of the Avon and the Stour the common lizard abounds, but the rarer sand-lizard is only known along the Ridgeway. The common snake is generally distributed, but the adder is confined to the oolite districts, such as Ilmington and the Edge Hills. The slow-worm occurs near Claverdon and Warwick. All three of the British newts exist in the county, but the palmated newt is confined to the south-west portion of it.

We have not space to allude to the land shells, the study of which, in spite of its interest, has been greatly neglected, but the spread of some alien species is a fact worthy of notice. One (*Dreissensia polymorpha*) has made itself at home in almost every canal and river, and another (*Physa heterostropha*), which comes from the United States, is fast establishing itself near Birmingham.

Warwickshire was in early times thickly covered with forest, and is well wooded at the present day, though in the north the woodland is less luxuriant. The trees in the valley of the Leam, the Arrow, and the Alne are conspicuous for their fine growth. The oak and ash are perhaps more abundant in the north of the county, and the elm in the south. The Warwickshire lanes have a special charm; they are often narrow, deep, and damp, and the variety of wild flowers which adorn them is very great. In some parts of Sutton Park are the remains of a past wild flora which cultivation has in other places extirpated, and in the woods near Warwick the lily of the valley is found. As there are no large lakes or fen land we cannot look for any special abundance of water plants, but at Olton reservoir is found the rare water-pepper,

Elatine hexandra, and in Sutton Park occurs the sedge *Carex Ehrhartiana*, one of the few places in Britain in which it exists. But perhaps the most interesting feature of Warwickshire botany is shown by the ponds at Chesterton, Itchington, Holt, and Southam Holt, the waters of all of which are brackish. Here occur plants of maritime affinities, *Rumex maritimus*, the golden sea-dock, *Scirpus maritimus* and *Scirpus tabernae-montanus*, and the celery.

Rookery in the Churchyard Elms, Stratford-on-Avon

Finally, turning to prehistoric ages, it must not be forgotten that Warwickshire is noteworthy among palaeontologists as first giving the name to the prehistoric amphibians known as Labyrinthodonts, found in the Keuper marl, and specially interesting for the toad-like footprints they left in the sandstone.

8. Climate.

The climate of a country or district is, briefly, the average weather of that district. This depends mainly on latitude, but also upon various other factors, all mutually interacting, such as the direction of its prevalent winds, its rainfall, the nature of its soil, its elevation, and its distance from the sea.

After latitude, the sea's influence is the most important, since it serves to soften down the extremes of cold and heat. These are always most marked in continental countries, and least marked in such tracts of land as happen to be surrounded by water. Continental countries are subject to colder winters and hotter summers than islands, and the latter are not only more equable but more moist.

The climate of Great Britain is less severe than other countries upon the same latitude, partly because it is an island, and partly from the warming influence of the prevalent south-westerly winds blowing from the Atlantic, which cause a movement of its surface waters towards our shores, bearing water appreciably warmer than the surrounding ocean. This comes to us from tropical America and washes along our western coasts, and is the chief reason why our winters are so remarkably mild.

Most of the weather we experience in these islands comes to us from the Atlantic, and it is of two kinds, settled or unsettled, anti-cyclonic or cyclonic. These terms require explanation. Any observant person standing

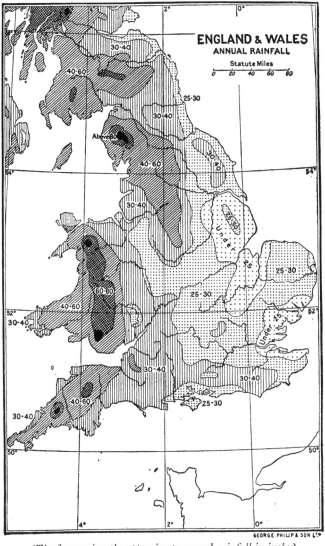

ENGLAND & WALES
ANNUAL RAINFALL

(The figures give the approximate annual rainfall in inches)

3—2

beside a stream may notice that while the main stream moves steadily on, the water at its side often flows in back currents, eddies, and small whirlpools. It is much the same with the air. The main current, which, as already stated, comes to us from the west and south, moves steadily forward, but it is fringed by small disturbed areas or eddies, and these develop on either side in the direction of the main stream. These eddies are the cyclones. But this band of cyclones does not always occupy the same position, it moves at times to the north, at other times to the south. Sometimes it passes clear of our islands altogether, at other times directly over them, but wherever it passes the weather is bad, disturbed, and unsettled. When a movement of the contrary nature sets in, it is called an anti-cyclone. The weather during its duration has a tendency to become fine or settled. Cyclones mean more or less storm, anti-cyclones are associated with calms.

The Atlantic ocean affects us in another way. It affects our rainfall. The air, moving over the vast extent of the ocean, becomes saturated with moisture, the water-laden clouds are driven across the mountainous districts of the west of England, and the moisture in them condenses and falls to the earth as rain. Naturally the amount of rain varies, but it decreases with remarkable regularity the further east we proceed, as is well shown by the map here given. The darkest portions are in the west, in Wales and Cornwall, in Cumberland and Westmorland, the lightest in the fens of Cambridgeshire and Lincolnshire, and round the mouth of the Thames.

But, as already stated, the climate is often affected by other factors, such as the character of the district, its position and its soil. The shelter afforded by a high range of hills, the point of the compass which its slopes face, the nature of its soil, whether a cold clay or a warm loam, all help to make differences in places sometimes but a small distance apart. The nature of the products of the soil must also be considered, a well-wooded district having a greater rainfall than a barren waste.

Careful records are kept in many places in England of everything relating to the weather, such as the amount of rain, the number of hours of sunlight, the rate of evaporation, the temperature, the direction and strength of the wind, and the amount of air-pressure. The principal Warwickshire station is at Edgbaston, under the control of the Midland Institute, and it is from their reports that the following details are chiefly taken, but there are over 30 other stations in various parts of the county, at which careful records are made daily. Warwickshire lies, as we see from the map, within that district which has an average rainfall varying from 25–30 inches annually.

The year 1912 was a very wet one, the rainfall measuring 8·71 inches in excess of the average rainfall of the county, as may be seen from the following.

> Average 1875–1900 28·98 inches
> 1900–1909 27·50 „
> 1912 37·69 „

And at Hampton-in-Arden it reached 10 inches above the county average, namely, 47·64 inches. The following table of the maximum and minimum rainfall over the five years from 1908 to 1912 will illustrate the degrees of variation within the county limits.

		Max.		Min.
1908	Birmingham . .	29·64	Barford .	21·40
1909	Farnborough . .	34·21	Coventry	24·52
1910	Tanworth . . .	42·11	Rugby .	26·05
1911	Tanworth . . .	27·50	Stratford	18·16
1912	Hampton-in-Arden	47·64	Rugby .	30·44

The greatest rainfall in these years occurred in the hilly districts of the "Midland Plateau," except in 1909 when it was exceeded in the Edgehill district. The lowest rainfall occurs in the valley of the Avon.

The influence of sunshine on the growth and perfection of all nature renders it most important that careful records should be kept of its daily duration. These are estimated by two systems, one in which photography, and the other the action of a burning glass, is made use of. The records here given were taken by the former system. In 1912 at the Edgbaston observatory, the lowest amount of sunshine was recorded in November, only 15·75 hours for the whole month. The highest amount was registered in April, 196·42 hours. The three summer months, June, July, and August, fell far below this figure, being 110·47, 123·33, and 97·75 respectively. The total amount of sunshine for the year was 1030 hours.

The wind direction, and its pressure in pounds on the square foot, are also noted, and by the automatic instruments which are used, observations are taken which enable continuous records to be preserved. The most noted gale recorded is that of the 24th of March, 1895, a gale which levelled many thousands of trees and wrecked the fine old cedars in the grounds of Warwick Castle.

The county is below the average in warmth, the mean temperature for the last 25 years being 46·7° Fahr. The temperature at Edgbaston varied in 1912 from 13·4° Fahr. in the shade on February 4 to 126·6° in the sun on June 22. In 1911 the thermometer reached 139·2° Fahr. in the sun on August 9.

9. Races, Dialect and Population.

We have no written record of the history of our land carrying us beyond the Roman invasion in 55 B.C., but we know that Man inhabited it for ages before this date. The art of writing being then unknown, the people of those days could leave us no account of their lives and occupations, and hence we term these times the Prehistoric period. But other things besides books can tell a story, and there has survived from their time a vast quantity of objects (which are daily being revealed by the plough of the farmer or the spade of the antiquary), such as the weapons and domestic implements they used, the huts and tombs and monuments they built, and the bones of the animals they lived on, which enable us to get a fairly accurate idea of the life of those days.

So infinitely remote are the times in which the earliest forerunners of our race flourished, that scientists have not ventured to date either their advent or how long each division in which they have arranged them lasted. It must therefore be understood that these divisions or Ages—of which we are now going to speak—have been adopted for convenience sake rather than with any aim at accuracy.

The periods have been named from the material of which the weapons and implements were at that time fashioned—the Palaeolithic or Old Stone Age; the Neolithic or Later Stone Age; the Bronze Age; and the Iron Age. But just as we find stone axes in use at the present day among savage tribes in remote islands, so it must be remembered that the weapons of one material were often in use in the next Age, or possibly even in a later one; that the Ages, in short, overlapped.

Let us now examine these periods more closely. First, the Palaeolithic or Old Stone Age. Man was now in his most primitive condition. He probably did not till the land or cultivate any kind of plant or keep any domestic animals. He lived on wild plants and roots and such wild animals as he could kill, the reindeer being then abundant in this country. He was largely a cave-dweller and probably used skins exclusively for clothing. He erected no monuments to his dead and built no huts. He could, however, shape flint implements with very great dexterity, though he had not as yet learnt either to grind or polish them. There is still some difference of opinion among authorities, but most agree that, though this may not have been the case in other countries, there was in our own land a vast gap of time between the people of this and the succeeding period. Palaeolithic man, who inhabited either scantily or not at all the parts north of

England and made his chief home in the more southern districts, disappeared altogether from the country, which was later re-peopled by Neolithic man.

Neolithic man was in every way in a much more advanced state of civilisation than his precursor. He tilled the land, bred stock, wore garments, built huts, made rude pottery, and erected remarkable monuments. He had, nevertheless, not yet discovered the use of the metals, and his implements and weapons were still made of stone or bone, though the former were often beautifully shaped and polished.

Between this Later Stone Age and the Bronze Age there was no gap, the one merging imperceptibly into the other. The discovery of the method of smelting the ores of copper and tin, and of mixing them, was doubtless a slow affair, and the bronze weapons must have been ages in supplanting those of stone, for lack of intercommunication at that time presented enormous difficulties to the spread of knowledge. Bronze Age man, in addition to fashioning beautiful weapons and implements, made good pottery, and buried his dead in circular barrows.

In due course of time man learnt how to smelt the ores of iron, and the Age of Bronze passed slowly into the Iron Age, which brings us into the period of written history, for the Romans found the inhabitants of Britain using implements of iron.

We may now pause for a moment to consider who these people were who inhabited our land in these far-off ages. Of Palaeolithic man we can say nothing. His successors, the people of the Later Stone Age, are believed to have been largely of Iberian stock; people, that is, from south-western Europe, who brought with them their knowledge of such primitive arts and crafts as were then discovered. How long they remained in undisturbed possession of our land we do not know, but they were later conquered or driven westward by a very different race of Celtic origin—the Goidels or Gaels, a tall, light-haired people, workers in bronze, whose descendants and language are to be found to-day in many parts of Scotland, Ireland, and the Isle of Man. Another Celtic people poured into the country about the fourth century B.C.—the Brythons or Britons, who in turn dispossessed the Gael, at all events so far as England and Wales are concerned. The Brythons were the first users of iron in our country.

The Romans, who first reached our shores in 55 B.C., held the land till about A.D. 410; but in spite of the length of their domination they do not seem to have left much mark on the people. After their departure, treading close on their heels, came the Saxons, Jutes, and Angles. But with these and with the incursions of the Danes and Irish we have left the uncertain region of the Prehistoric Age for the surer ground of History.

The Brythons had been settled in Warwickshire some centuries when the Romans conquered it, and, according to Professor Rhys, were of the tribe called Dobunni, who were hardly independent of the more important Catuvelauni. In process of time the country became gradually Romanised, and all persons of any in-fluence adopted the Latin language as their own. This

is the reason why so small a part of modern English consists of Brythonic words.   In Warwickshire some river
names, such as the Avon and Alne, and a few place-
names, may have such an origin, or they may be even
earlier still.

This part of England was conquered by Teutonic
tribes at a late date, when they were themselves beginning to lose their earlier vigour.   The part now called
Warwickshire, or at least the southern part of it, was
largely occupied by a tribe of West Saxons calling themselves Hwiccas, but they in turn had to give way to the
Mercian influence, and in the far north of the county
there are traces of East Anglian conquest.

During this period three main dialects were spoken in
England.   In the north, Northumbrian, the parent of
Scotch; in the middle Mercian; and in the south-west
Saxon.   The Warwickshire man was able to understand
something, at least, of the dialects north and south of
him, and in the southern part of the county West Saxon
was spoken for a considerable time.   In the course of
time the East-Midland speech drove the others from the
field, principally because it was the language of London.
With the advent of the Normans their scribes re-spelt
the whole, and introduced many French words, which
were added to very considerably in the French wars.
The whole gradually settled down into the "Middle
English" of Chaucer and Hampole.

The Danes, although close to our county, do not
appear to have influenced it much.   They kept to the
north of the Watling Street.   The later Danish invasion

of Cnut passed through the district but there was little permanent settlement. At the time of Domesday Book small groups of foreigners had made their home here, among them families of French and Flemings. There are also traces of Welsh and perhaps of Scots. At a later date Welshmen and Irish were constantly passing through

Morris Dancers at Ilmington

the county, and the trades of Coventry may well have given employment to artificers from abroad, who might be encouraged to settle and introduce improvements. In the sixteenth century Flemish tapestry-weavers were plying their craft in the village of Barcheston.

For these reasons Warwickshire folk may be expected to retain evidences of a mixed descent. There are considerable numbers of a small dark race, the descendants doubtless of the Romano-British farmers, who were probably neither exterminated nor exiled. Intermarriages may have taken place and resulted in the medium race, who are neither of the fair tall Teuton type, nor of the Celtic dark race.

The Normans did not replace the Saxon landowners in the county. The bishop, a Saxon, retained his old tenants, as did those in authority at the great priory of Coventry, and even where the tenant-in-chief was a Norman baron the actual holder of the land was frequently a Saxon.

The traditions of the county and its folklore retain much of the past, especially in the villages bordering on the Cotswolds; in some of these Christmas Mummers are well remembered and Morris dancing is still practised.

10. Agriculture.

Warwickshire, apart from its manufacturing districts, is entirely an agricultural county. It must not be forgotten, however, that within living memory most villagers used their spare time in the production of home-made goods. Home-spun flax produced the clothes of the people as far as linen went. Shag, a coarse species of plush, was largely produced at Brailes, linsey-woolsey was made at Shottery. Bearley had in the eighteenth century a flourishing paper mill, and Halford was famous for rush-seated chairs and workmen's baskets. There was a fulling mill at Binton, and a cutling mill at Atherstone-upon-Stour, and tapestry works at Barcheston, of which more will be said. It is interesting to note that very successful work of this description is now being carried on at Shottery. Needle-making spread from Redditch to the villages round, and ruins of works can still be seen at Morton Bagot and elsewhere. Glove-making was also carried on in the country districts. In spite of all these small industries the county was, and is, as just stated, mainly agricultural.

Early writers divide Warwickshire into the Feldon and the Weldon, the latter lying to the north of the Avon, the former to the south. These names have little more than a literary value, but they represent roughly the character of the two portions of the county, the Weldon being better wooded. It must not be forgotten that until the eighteenth century most of the country-side lay in strips and balks, very much like modern allotments. The only place where they still appear is at Crimscote in the parish of Whitchurch. In this open field system a third of the land lay fallow every year, the remainder being cultivated. This was known as the three-course system. The first year wheat or rye was grown, the second barley, oats, beans, or peas, the third year the land rested. Neither turnips nor artificial grasses, it must be remembered, were cultivated in the Middle Ages. In fortunate places there were also some small pastures and more rarely meadows, but where there was neither the cattle and sheep were depastured on the waste and fallow. With the introduction of turnips the system developed into a four-course one, the roots being sown the first year. By the end of the eighteenth century much improvement had been effected, in which Joseph Elkington of Princethorpe played an important part. The system then followed was a five-course one: (1) turnips, (2) barley, (3) oats, (4) clover, (5) wheat, or oats repeated in the place of wheat. In the middle of last century a six-course

Shottery Peasant using a Breast-plough

system was in vogue: (1) turnips, (2) barley, (3) seed or peas or beans, (4) wheat, (5) beans, (6) wheat.

The ploughs used in the county differed but little from those used elsewhere; a curious and cumbrous tool, the breast plough, is still occasionally used but will soon be entirely forgotten; the illustration of a Shottery man using his breast plough shows the method employed.

Sheep farming was considered profitable at an early date. It was introduced partly on account of the dearth of labourers caused by the several visitations of the Black Death, and partly owing to the improved character of English wool. Its introduction accounts for some, at least, of the depopulated villages of the county. In the

Warwickshire Bullock

eighteenth century the sheep were of two classes : (1) short-woolled or field sheep, (2) long-woolled or pasture sheep, a variety being known as the "Warwickshire breed." It is described by some writers as "intolerably bad."

The cattle, by which most of the farm work, such as ploughing, cartage, etc., was at one time done, developed from the aboriginal short-horned race into a long-horned breed of considerable merit, which is still in favour, and has a class to itself in the county agricultural shows. At the close of the eighteenth century Warwickshire was almost entirely a grazing county, with dairies producing much cheese. The continental war and the high price of cereals altered it to a corn-growing one, but it is now rapidly turning back to pasture, though butter has largely taken the place of cheese-making, while great quantities of milk are sent to London, Birmingham, and Coventry.

The cart-horses used in the county were the old Lincolnshire breed, and the chief horse fair was held at Rugby. Pigs were much bred, the Berkshire being a general favourite, except in the Tamworth district, which had then, as now, an excellent local variety.

After the Warwickshire farmers gave up breeding the local long-woolled breed of sheep, other and better breeds were introduced. In the Alcester district Leicesters crossed with Shropshires were the favourite, while further south North Cotswolds crossed with Shropshires and Southdowns prevailed. In some of the parks foreign breeds may be seen. Thus Arbury has some African sheep, and Charlecote a flock said to have originated in Spain.

The open-field system passed away between 1760–1830. With the enclosure of the land new crops, new tools, and new systems came in. Mangolds and swedes and prickly-comfrey were grown, and artificial foods and cake introduced. The repeal of the Corn Laws later led to still further changes. The price of wheat fell, but the introduction of machinery helped to balance matters, and from that date until 1874 the county was prosperous. The most disastrous year was in 1879, when one-twelfth of the farms in the county were to let.

11.   Coal Mining and Quarrying.

The county of Warwick owes its great industrial development very largely to the important coalfield which underlies much of its northern area. The series of sandstones and shales in which the coal-seams are found have a thickness of about 1000 feet, and lie in a basin with its longest axis running north and south in the district between Tamworth, Coventry, and Nuneaton. There are five seams, but the " Four-foot " is not worked. The workable seams are known as the "Two-yard," which as its name implies is six feet in thickness; Base coal, two feet; Rider coal, four feet; and Ell coal, four feet. The industry employs nearly 17,000 persons.

Comparatively little is known of the earliest history of the field, but coal-pits existed at Griff and at Chilvers Coton in the thirteenth century, and there are records extant showing that coal of considerable value was raised on land belonging to the abbey of Nuneaton in the century following, though these early pits were necessarily of small size. The use of coal must, however, have been very limited owing to the difficulty of transport.

By 1600 the number of coal-pits had largely increased, and the methods of dealing with the water in them had improved. Headings drained the water by means of wooden drains into a water-pit from which it was conveyed to the surface by means of a gin worked by horses or mules. The mines at Griff at this period did not produce any very good return for the labour, the profits rarely exceeding eight per cent. This is no doubt partly the reason why they were frequently standing idle. Another cause was the opening of pits at Bedworth. The quarrels between the rival coal-owners, taken into the Law Court, fortunately throw considerable light on these early mining operations.

By the first quarter of the eighteenth century there were some 50 collieries along the edge of the Warwickshire coal-field, that at Griff being "40 ells in depth and of vast compass." About this time steam-engines superseded the old horse-gin, and engines were at length constructed with cylinders of great size. The water was got out much more rapidly and a greater depth reached.

Another century passed before anything was done to provide artificial ventilation, but the Warwickshire coal was fortunately free from gas, and although from its nature spontaneous combustion occurred from time to time, explosions were few. The most fatal on record took place at Baddesley, May 2, 1882, when 22 lives were lost. Modern mines, owing to artificial ventilation, can be worked at very great depths, thus Griff is 750 feet, Exhall 855, Ansley 1140, and the Charity Colliery 1200 feet. The use of electric light and coal cutting machinery is also an important improvement.

The increase in the output of late years has been

enormous. While in 1860 only 545,000 tons were raised
and in 1870 hardly more, the output for 1900 exceeded
3,000,000 tons, and in 1911 was 4,893,483 tons.

Other minerals of commercial value are produced
from coal mines such as fire-clay and iron pyrites, 26,894
tons of the former and 5650 of the latter were raised in
1911. There are extensive quarries in other parts of the

Kingsbury Collieries

county. Ironstone was formerly worked at Burton Dassett, though but little now. Warwickshire has always
been productive of good building stone, as witness the
houses of yellow Lias in the Edgehill district. There
were ancient quarries belonging to the Crown at Coventry,
and the Red Sandstone of Kenilworth was much used by
the monks. Attleborough and Rowington still produce
good stone. The Lower Lias was worked at Binton,

Grafton, Rugby, and Ilmington, and Wilmcote had a flourishing cement factory, which has lately closed down. Cement works on a larger scale, however, exist at Harbury, Southam, New Bilton and Ettington.

Road metal is obtained from the Cambrian rocks of Tuttle Hill and Oldbury, and the diorite of Chilvers Coton and the hard limestones of Ratley are used for

Diorite Quarry, Chilvers Coton

a similar purpose. Gypsum, much used at one time for cheese-room floors, was obtained until recently in the neighbourhood of Spernal.

At Nuneaton is an extensive pottery for the manufacture of drain-pipes, tiles, vases, and chimney-pots. The material used is a series of coloured marls dug from pits adjoining the works. At Stoke a coarse brown ware is made, and china door-knobs at Birmingham.

12. Industries.

In the Middle Ages the commerce and industries of towns and cities was largely under the management of guilds. These societies consisted of men and women who combined for trade as well as for religious and social purposes. Membership was not confined to the locality, indeed the further afield the guild members spread, the greater the advantage to the body. These guilds were not only concerned with the petty details of trade, but even with such matters as the policing and scavenging of towns, and protection against fire. There were powerful guilds at Coventry, Warwick, and Stratford-on-Avon ; and even Birmingham, though not a corporate town, had a guild of considerable influence.

The metal trade of Birmingham is enormous, but it is probably not so ancient as that of Coventry, where cutlers were working at their craft as early as the first quarter of the thirteenth century. Cutting tools were turned out at Birmingham in 1558. Less than a hundred years later sword-blades were made there ; a sword-maker, one Robert Porter, made 15,000 swords for the Parliamentary army, a proceeding which brought trouble to the town. At the end of the eighteenth century there were four firms in the trade, and Government orders were obtained and large consignments were also sent to Mexico, Brazil, and elsewhere.

The gun trade of Birmingham dates from 1683. It was obtained for the town through the influence of

Sir Richard Newdigate of Arbury. It later became a famous trade and a serious rival to that of London, and over a million and a half stand of arms were manufactured here between 1814 and 1817. The gunmakers worked at high pressure during the Crimean War, and received extraordinary wages for overtime. The American Civil

Body Shop, B.S.A. Works

War was also largely beneficial to them. At the present time the Birmingham Small Arms Company manufacture bicycle parts in addition to weapons. Flint-lock muskets were made for African trade purposes as early as 1698. Proof-houses for testing gun-barrels were erected in the town, and such barrels as passed the test were stamped with the Birmingham mark.

In the eighteenth and nineteenth centuries the town bore an important part in minting. The famous Soho works of Matthew Boulton and Watt struck copper coins for the East India Company, and also for the British Government in 1796 and 1806, as well as for Russia, Denmark, Spain, and other foreign countries. The Birmingham coinage is distinguished by a letter H.

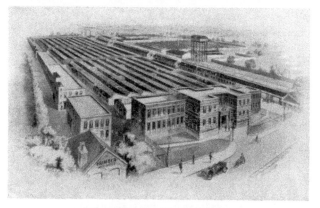

Humber Works, Coventry

Copper tokens were largely made to supply the lack of copper coins during the eighteenth and early nineteenth centuries. The Birmingham mint is still at work, and many foreign countries obtain their coinage from the city.

Machinery of all kinds has been produced both at Birmingham and Coventry. The Cornish mines were thus supplied with "fire engines," and castings and engines of all descriptions were made at the Soho works from

1775 to 1861. Railway carriages were first built here
in 1838, and nearer our own time Coventry turned out
vast numbers of sewing machines, and still later of bicycles,
and the various famous cycle firms have now turned their
attention to the production of motor cars. Of this trade
Coventry is the English centre; and the Daimler Company produced their first car here in 1896.

Bell-founding did not begin in Birmingham until the
middle of the eighteenth century, but there were founders
working at Warwick in early medieval times, and their
bells remain at Beaudesert, Halford, and Whitchurch.
Bells were also cast at Stratford by Worcestershire men.
Brass-ware is also a noted Birmingham production, and
dates from 1740. Brass bedsteads were invented by a
Birmingham man, Benjamin Cook, and in 1885 nearly
40,000 tons of bedsteads were sent from the city. Iron
was much wrought in Birmingham from its earliest days,
and all descriptions of smaller iron goods are produced in
great quantities.

There was a guild of wiredrawers at Coventry as
early as 1430. Machinery was introduced into the trade
in 1565, but in recent years the manufacture of brass and
copper wire (and also of steel wire) has been carried on at
Birmingham. The pewter trade of the city was a much
later affair and only began in the middle of the eighteenth
century; large quantities of candle-moulds, pewter plates,
and dishes being then turned out. Pewter was afterwards
superseded to a great extent by the discovery of Britannia
metal, the manufacture of which came to the town in
1814 from Sheffield.

Warwickshire has been from a very remote date the home of the needle and pin industry, the modern centre of which, though actually at Redditch in Worcestershire, is on the very edge of our county. Studley, however, was in 1700 the chief seat of the industry, the waters of the Arrow turning many mills. Here a horse-power mill was introduced in 1701. Afterwards the mills round Studley were superseded by those of Alcester, where sail-needles and packing-needles were also made. It was not until 1828 that machinery began to take the place of hand labour, with the usual result of much opposition. In 1835 Henley-in-Arden had a factory producing not only needles but fish-hooks. Redditch overflows into Ipsley parish, and our largest and most important firm of needle makers has its main works in Warwickshire. Pins are chiefly made at Birmingham.

The earlier commercial history of Birmingham is largely concerned with the button-making industry, but it was also carried on at Stratford and at Coventry as early as the time of Elizabeth, buttons of almost every kind of material from gold and silver to horn being made. Buckles, at a time when shoe-buckles were a fashionable necessity, were manufactured in vast numbers in the town. The patterns were various, but the general shapes were either round, octagonal, or oblong. Buckles went out of fashion by the introduction of the shoestring, and by 1823 the 27 button-makers in Birmingham had been reduced to two.

Clocks were not used in houses until about 1615, and soon afterwards the trade was introduced to Coventry, and watch-making being combined with it, it rose to be the staple trade of the old Warwickshire city. When Edward Loxley invented the compensation balance in 1857 there were 2000 watchmakers in the city with 3000 apprentices. Watches are also made in Birmingham, where the silver and jewellery industry has now reached very large proportions.

The manufacture of steel pens was introduced into Birmingham by Samuel Harrison, who made the first steel pen in the city for Dr Priestley. Handmade pens were manufactured here by James Perry as late as 1820, in which year Joseph Gillott invented and made the three-slit pen and adapted machinery for its production. In 1876, 3000 varieties were produced by one firm alone for its customers; in 1887, 160,000 gross per week were turned out, chiefly by female labour.

In several of the Warwickshire towns the tanning of leather used to be carried on. There was a Tanners Street at Stratford at an early date, and Coventry had its Leather Hall; there were also regulations under the guild of Skinners of Warwick, who met there in their Hall. Allied to this was the saddlery trade, which had its headquarters at Birmingham, and still employs a considerable number of workers. The glove trade was carried on in Coventry as early as 1434, and afterwards flourished at Stratford in the reigns of Elizabeth and the Stuarts. The wares were sold at the High Cross, but the trade never had the repute in Warwickshire that it held in Worcestershire.

An important local Coventry trade in early days was

in blue thread, though its memory survives only in the proverb "True as Coventry blue." The thread fetched six shillings per lb. in the reign of Elizabeth.

Glass was also made at Coventry as early as 1302, but it was not an established trade until the eighteenth century, when it had Birmingham for a centre. One of the best known English glass-painters was nevertheless a Coventry man, John Thornton, who was employed to glaze the great east window in York Minster. There was a glasshouse at Coventry in 1696, others existed at Birmingham in the eighteenth century. At this town glass was extensively used for making buttons. Here too Francis Eginton (1736–1805), whose work may still be seen in St George's Chapel in Windsor Castle, painted windows at the Soho works. He also produced paintings which he called "polygraphica," but they ceased to be made after 1780, and the secret was lost, although his son carried on the business. In more recent times stained windows have been made by John Hardman and William Holland.

Cloth-making is a widespread and very ancient industry. As early as the fourteenth century there was a powerful weavers' guild in Coventry, and a fullers' guild and fulling mills. There was also later a clothiers' company, the members of which produced the various camlets, tammies, and other local materials called "Coventry ware." Ribbon-weaving was one of the most important of all the Coventry trades. It was introduced in the early days of the eighteenth century, employed a large number of hands, and flourished until a taste for French ribbons set in. Its decay was hastened, however, not only by the change of fashion, but on account of a series of disastrous strikes. There was a brief revival in 1870.

The making of caps and hats was another early Coventry industry. In Tudor times there were many disputes among the cappers of the city, and Acts of Parliament were passed between 1488 and 1563 to regulate them. In Elizabeth's reign felt hats were introduced The hatters of to-day carry on their manufacture at Atherstone, Nuneaton, and Rugby.

It must not be forgotten that trade in Coventry has adapted itself wonderfully to altered circumstances. The failure of one manufacture has led to the birth of a more successful one; the city is moving forward with rapid strides and the staple industry of to-day is that of motor cars.

13. History.

The history of the county begins with the Roman period, when the conquest of the Midlands was carried out by the generals of the Emperor Claudius in A.D. 43–47. In this short time the Roman legions had reached Shrewsbury, but, apart from the roads, few traces of their passage remain in what we now term Warwickshire. In all likelihood there was little fighting, for the country seems to have been but thinly populated. The district soon settled down, and remained peaceable during the whole of the Roman occupation of these islands.

The Romans left Britain gradually after 410, and for the next two hundred years the Midlands remained in the hands of the Cymry, as the native Celtic population were called, and much of the Warwickshire of that date may have been forest, though its traditional name of Arden is somewhat doubtful. It was at length invaded from the south-west and overrun by a Saxon tribe called Hwicca, an offshoot of the conquerors of Wessex. This people appears to have invaded the valley of the Avon, advancing partly by the river, and partly along the Fosse-way. Their conquests extended as far as south Warwickshire, Worcestershire, and part of Gloucestershire, and by 679 they had a diocese and bishop of their own.

The north of the county was conquered by a tribe of Angles under the chieftainship of Cridda, who became first king of Mercia in 586. The Mercians gradually brought the Hwiccas under their authority, but allowed them under-kings or princes of their own. Penda, King of Mercia, was the last vigorous upholder of the old heathen religion, his son Peada became a Christian, and in 656 formed the see of Lichfield, the first bishop being Duima, a Scottish monk from Lindisfarne. It was his more famous successor, St Chad, who made Lichfield his cathedral town.

The capital of Mercia was at Tamworth, on the extreme edge of the county, and here several events of interest took place. In the meantime a new race, the Danes, were fighting their way southwards, leaving behind them burnt-out villages and slaughtered Mercians. They were opposed successfully by a very remarkable

woman, Aethelfleda, "Lady of the Mercians," who drove
them from her territory, and built a chain of forts across
it, one being at Tamworth, another at Warwick, which
has ever since been the county capital. Warwickshire
then had peace until the second Danish invasion under
Cnut in 1016, when it was sorely harried, and its religious
houses burned. The old kingdom of Mercia had passed
away long before, to become part of a united England,
and its kings were replaced by Earls, the most famous of
these being Leofric and his Countess Godgifu or Godiva.
This pair founded the great abbey of Coventry in 1043.

The Norman Conquest had less effect upon the Saxon
population of Warwickshire than elsewhere, and one
large landowner, Thurkitill, continued to hold his vast
possessions until he died. The Norman bishop of Worcester, Wulstan, and the Saxon abbot of Coventry, held
their lands in peace, and a very large number of under-
lords remained undisturbed. The spirit of unrest was
nevertheless in evidence, and William himself came to
Warwick in 1068, and repaired its earthen walls, creating
here a Norman Earldom with a seat at the castle.

The general scheme of defence after the Conquest
was planned to resist invasion from the north and west,
so that a series of strong castles and great monasteries
barred any advance on the capital ; in Worcestershire
the great monasteries were those of Worcester, Malvern,
Pershore, and Evesham, in Warwickshire castles were
more prominent, viz. at Tamworth, Kenilworth, and
Warwick. There was also the walled city of Coventry
and its castle. The civil wars of Stephen and Maud

showed some of the weakness of the scheme. The strongholds were garrisoned by Robert Marmion of Tamworth on the one side, and by the Earls of Chester and Warwick on the other. Towards the end Stephen seized Warwick Castle, but it was easily recaptured by the widowed Countess Gundrada, who held it for the Empress.

No great event happened in the county during the reigns of Richard and John, but it played an important part in the strife between Henry III and his barons. Kenilworth was the centre, and it was here, in the stronghold of the Clintons, that De Montfort imprisoned the King and his son in the year 1265. The prince, however, contrived to escape, and gathering an army together surprised both the Earl and his troops, who were lodging in the priory and town. Earl Simon fled half naked, and his men were scattered. The prince then retired to Worcester, but contrived to throw himself between the armies of the Earl and his son, preventing their junction, and defeating and killing the former at the decisive battle of Evesham. This was followed by a long siege of Kenilworth Castle, which lasted for six months, with the whole fighting power of the country assembled before its walls. During its progress a Parliament was held, but the "Dictum de Kenilworth" then drawn up failed to secure the surrender of the castle, which was at last subdued by famine.

In the following reign, that of Edward II, affairs of great importance happened in Warwickshire. The King and his barons were usually quarrelling, the King choosing unwise and weak favourites, the most foolish being Peter of Gascony (Piers Gaveston) who in an unguarded moment sneered at the haughty Earl of Warwick, calling him the "dog of Arden." Unfortunately for Gaveston he was captured by the Earl, and tried by his peers in Warwick Castle. An unrepealed decree enabled him to be legally executed. The execution was hurriedly carried out on Blacklow Hill in 1312, where a monument commemorates the deed. The King could not at the moment retaliate, but later the Earl of Warwick had to retire from Court. In 1326 the King was taken prisoner by Henry of Lancaster and brought to Coventry, where he was forced to abdicate, and shortly after perished.

The Black Death wrought immense havoc in the county during its three visitations. It had far-reaching social results, which led to the various movements among the peasants ending in the rising under Wat Tyler, one of whose principal assistants, John Ball, attempted to conceal himself in Coventry, but was captured there and executed. In the year 1404 another Parliament met in Coventry and made an attempt to tax the Church. It was, however, easily defeated by the Archbishop. It was in this city that Henry V received the Dauphin's insulting present of tennis balls, which led to the invasion of France, and the English victory at Agincourt.

Although no battle in the Wars of the Roses was actually fought in Warwickshire, yet Warwick's Earl, Richard Neville (the King-maker), made our county the centre of many plots and counter-plots. The Earl himself was often of far greater power and influence than the

King. In spite of this he was attainted, together with the Duke of York, by the Parliament held at Coventry in 1458–9. The Earl fled abroad, but soon returned while the Court was still within the city. He landed in Kent, marched on London, only to be defeated by the Lancastrian army at the battle of St Albans. Edward, nevertheless, began to mistrust his powerful supporter, and by suspicion and slights drove him to act on the side of Henry. Edward himself was deserted by his own men, and captured at Honiley, from which place he was taken to Kenilworth, and thence to Coventry, where he perhaps witnessed the execution of Earl Rivers and his son on Gosford Green.

Early in 1470 Edward was master again, and the Earl of Warwick fled, but he soon returned as champion of Henry, who again became King, but it was not for long. Edward crossed the sea, and the Duke of Clarence played a traitor's part, joining in the attack on his father-in-law the King-maker, who fell fighting on the field of Barnet. Thereafter, little of historical interest happened in the county until the foolish attempt of the Duke of Suffolk to dethrone Queen Mary. The Duke fled to Astley Castle, and hid inside the trunk of a hollow oak, but was betrayed by one of his own men and beheaded.

In the reign of Elizabeth Warwickshire again plays a conspicuous part in English history. The "good Earl" (Ambrose Dudley) held Warwick Castle, Elizabeth's favourite the Earl of Leicester owned Kenilworth, and the Queen was often in the county. Mary Queen of

Warwick Castle

5—2

Scots was brought a prisoner to Coventry in November, 1569. Elizabeth was magnificently entertained at Kenilworth by the Earl of Leicester in 1575, and all the time of her visit masques and other entertainments were lavishly indulged in for her amusement. It is possible that Shakespeare may have witnessed some of these.

In the early years of James I, Warwickshire became the centre of the plot of the Gunpowder conspirators, most of whom were disappointed local men. A hunting party had been arranged by them on Dunsmore Heath, and here it was that they first learnt the news of its failure. The conspirators were then chased through the county into Worcestershire and those that survived their capture met their death on the scaffold.

The early stages of the fight between Charles I and his Parliament were all fought out in Warwickshire. In 1640 Warwick Castle was garrisoned by Lord Brooke, while the King garrisoned Kenilworth and Tamworth. Birmingham threw itself with enthusiasm into the conflict, and forged sword-blades for the soldiers of the Parliament. Three weeks before war was formally declared Lord Brooke and Lord Northampton met and fought. The siege of Warwick Castle followed, but the Royalists failed to capture it. The King, then at Stoneleigh, marched to Coventry, but was refused admission for his army, although the citizens offered to receive their sovereign and a small guard. Charles, very angry, turned his cannon on the Newgate, but the fears of the people were relieved by Lord Brooke's approach to relieve them. A number of skirmishes then ensued,

but the first serious battle of the war was fought at
Edgehill, October 22nd, 1642. The result was indecisive,
Prince Rupert and Lord Wilmot defeated the cavalry
of Essex, but their reckless pursuit laid open the centre
and the right to the enemy's infantry.   The King's

Ratley Round House, Edge Hill

(*The King, according to tradition, took up his position here*)

standard was taken, but recaptured, and Lord Lindsay
fell.   Nevertheless Essex had to withdraw his troops to
Coventry, and the King was able to proceed on his
march towards London.   Many Warwickshire gentlemen
joined him, but there was no longer any chance of

holding the county.   Prince Rupert, however, raided Birmingham, and nearly burnt it to the ground.

The romantic escape of Charles II after the battle
of Worcester was made in part through Warwickshire,
and a lane near Stratford is still called King Charles's Lane
in memory of the event.   When the King "came to his
own again" Coventry endeavoured to avert the memory
of its misdeeds by unusual rejoicing, but the Earl of
Northampton received orders to demolish its defences.
"Never more could Coventry be a city of refuge for dis-
tressed Queens, nor could it refuse entry to a King at
the head of his army."

From this time the history of the county concerns
itself alone.   It is a tale of occasional riots and election
disputes combined with immense industrial developments.

14. Antiquities.

We have already, in chapter 9, spoken of the
earlier inhabitants of Britain, of Palaeolithic and Neolithic
man, and of their successors of the Bronze and Iron Ages.
Palaeolithic implements have been found at Saltley,
Wellesbourne, and Walton, and careful search would no
doubt lead to discoveries elsewhere.   Weapons of Neolithic man have often been discovered, especially towards
the south and south-west of the county, where there are
camps at Ilmington (Foxcot), Nadbury, and Henley-in-
Arden.

A few of the rivers in the county may have been named by these people, but we have no knowledge of their language, to judge if this is so.

A very long time afterwards that branch of the Celtic race whom we know as Goidels came over from the continent, armed with weapons of bronze, so that they were able to drive out the older race. These people also left their weapons behind them, among others a very beautiful dagger, which was found at Bilton near Rugby. They usually burnt their dead, and piled up a conical mound of earth over the urn containing the ashes. They were great farmers, and the introduction of bronze enabled them to make sickles with which to mow their corn. They sometimes built themselves huts, made much better pottery than their predecessors, and ornamented it with patterns drawn on the clay before baking. They landed here about 1500 B.C., and were driven to the west about 600 years later by another branch of the same race called the Brythons. These probably obtained the mastery because they knew the use of iron. The Brythons were in turn conquered by the Romans in A.D. 43. They were not driven out, however, but became gradually Romanised, losing their own language and customs, and copying those of their conquerors.

What little there is to say about Roman Warwickshire has already been said ; there is hardly any relic of importance known to have been found in the county. A kiln at Hartshill, another at Brailes, a number of brooches and ornaments at Alveston, a coffin at Alcester, and a few fragments of tiles and pottery, together with large numbers of coins, constitute the chief items. A very perfect penannular brooch has recently been found at Stratford and is an excellent example of its type. These

British penannular Brooch

brooches were nearly circular and had a long pin. Of early Saxon remains there are more, and some of these happen to be of particular interest. The course of the Teuton invasion can be traced along the valley of the

Avon and the Fosse-way by their burial places. The richest graves yet found were discovered at Longbridge in 1875; weapons, brooches, armlets, remains of bronze vessels, and a remarkable gold disc on which a scene from the Sigurd Legend is engraved, form part of this treasure. Another gold disc was found at Compton Verney, and important hoards were also discovered at Cestersover and Oldbury. Still more interesting was the carved walrus-ivory crosier-head found at Alcester, one of the most beautiful works of Saxon art ever found in England. The Saxons left behind them camps of a peculiar form. They consisted of a high mound with an oval base court attached. We know that the " Lady of the Mercians " constructed earth forts at Tamworth and Warwick, but there are many more in the county of very similar design, as at Brandon and Brinklow, Morton Bagot, Oldberrow, and Ratley. Many modern authorities believe these to be Norman, but the proofs do not seem to be conclusive, and the Saxon land-owners must have used some kind of simple defences.

15. Architecture—(a) Ecclesiastical.

Architecture as an art may be said to begin in England some time before the Norman Conquest. The builders, accustomed only to work in wood, had gradually to learn by experience how to construct edifices of stone. They learnt at first very slowly, and their early efforts were rude enough. Nothing was known of strain and thrust, so they built no buttresses and their churches stood by mere massiveness. The walls are often marked by an

Shaft of Cross, Kinwarton

arrangement resembling that of a fish-bone, called herring-bone work, and by stones placed alternately vertically and horizontally in the quoins or corners of their towers,

known as "long-and-short work."   The sculpture of
this early date is very rough, though interesting.   It
consists of interlaced patterns, such as are found in the
cross shaft of the tenth century at Kinwarton, or on
the early *tympana* at Alveston Old Church, or figures
such as the curious Agnus Dei at Whitchurch, or the
St Christopher built into the wall at Billesley.   There

Oxhill Font

(*The base is modern work*)

is also of this period a very remarkable font in Oxhill
Church.   The best example of an early church of this
date in the county is that of Wootton Wawen, where
much of the tower and part of the nave walls are
pre-Norman.   Other churches having early work of this
kind are Haselor, Offchurch, and Loxley.

   With the advent of the Normans matters improved;
slight buttresses were used, but more for ornament than

support. The principle of the vault was introduced, and round-headed doors and windows, so planned that one arch is recessed within another and often richly ornamented with chevron and other mouldings. Interesting churches in the Norman or Romanesque style as it is termed, are found at Halford, Barton-on-the-Heath, Beaudesert,

Norman Chancel Arch, Beaudesert

and Oxhill, and at Kenilworth, where a very peculiar and handsome doorway remains. Pre-Norman and Norman features may be seen here and there in the smaller churches all over the county, but they are richer in the south, the finest of all perhaps being the aisled church at Tysoe, and at that of Burton Dassett.

Most of Warwickshire, as already mentioned, was within the diocese of Worcester, its cathedral being in that city, but Coventry had a cathedral, of which, however, scarcely any trace remains. Moreover, there were no religious houses in the county that could at all compare with the great foundations of Gloucestershire and Worcestershire. The abbey of Coventry has gone with its cathedral, and but little is left of the priory of Kenilworth. Merevale Abbey is but a ruin, though it still retains features of great interest, among them a refectory pulpit. There are some remains at Stoneleigh, and others at Studley and Pinley.

These religious foundations were the great civilising power of their day, for within their walls men were able to study and copy the writings of the past. Art was encouraged and practised, and science also, so far as it was understood. Their wide lands were models of agriculture, and the majority of the inmates did their duty faithfully to God and their fellow men, receiving all who demanded food and shelter, lodging great merchants, nobles, and kings. They played a very great though quiet part in the development of England. The Benedictines possessed houses at Coventry, Alcester, Alvecote, Polesworth, Henwood, Nuneaton, and Wroxall ; the Cistercians at Combe, Merevale, Stoneleigh, and Pinley ; the Carthusians at Coventry ; the Austin Canons at Kenilworth, Arbury, Maxstoke, Studley, and Warwick ; the Austin Friars a small cell at Atherstone ; the Knights Templars had a preceptory of their order at Balsall, and granges at Warwick and Temple Grafton ; while the

Dominican Friars had houses at Warwick; the Franciscans and Carmelites at Coventry; and the Trinitarians at Thelesford. There were several hospitals, some of which became famous, that of Spon in Coventry, of St Thomas in Birmingham, of the Holy Cross at Stratford, and the leper house of St Michael in Warwick. There were

Crypt, St Mary's Church, Warwick

also several collegiate churches at Astley, Coventry, Knowle, and Stratford, and—most important of all— that of St Mary, Warwick.

With these monastic churches to serve them as models, the builders of the country churches set to work, using such material as came to hand, either the oolite of the Cotswolds, the red sandstone of Kenilworth, or the lias and sandstones, grey or brown, of other parts. The Warwickshire churches show that they were largely designed by local men, and carried out with local materials. In one church, Kinwarton, there is a window with wooden tracery, a peculiar and graceful feature.

The passage from the massive Norman style to the light and graceful Gothic was a very gradual one, which occupied most of the last quarter of the twelfth century. While in progress, we find pointed arches arising from Norman capitals, as at Whitchurch, and round-headed windows in the same walling as pointed doors. It was a time of great activity in church building, but passed almost imperceptibly into the graceful purity of the first Gothic style usually called Early English. The masons had by this time learned to economise material by throwing the lines of pressure on certain points, which were supported by buttresses. It was a great discovery, and led to the beautiful buildings of our great abbeys and cathedrals with their flying buttresses, carrying the weight of heavy roofs over the side aisles to the ground. The Early English style is marked by its wonderful lights and shadows, its deep-cut roll mouldings, usually with a narrow fillet on the edge of the roll, its acutely-pointed narrow windows, and arches springing from slender columns, which are often detached from the main wall. Their carvings are vigorous and natural in place of the conventional lines of the Norman work, and the roofs are high pitched. The church of Priors Hardwick is almost entirely in this style, as is that at Pillerton, and portions of that at Stratford-on-Avon.

Towards the close of the thirteenth century, the windows, which had been often grouped in triplets, began to exhibit tracery. There are a few beautiful examples left, but they are not common. This new development opened the way for the second Gothic style, the

Holy Trinity Church, Stratford-on-Avon

Decorated, which began comparatively simply, but at last reached an extravagance of ornament which was only put an end to by the cessation of building at the Black Death in 1348. The most striking feature of Decorated churches are their lofty spires. The mouldings are more shallow than in Early English. The arches are broader,

St Michael's Church, Coventry

the window tracery elaborate, and often arranged in geometrical patterns. There are numbers of churches in this style in the county. That of Stratford was rebuilt at this date, and possesses some excellent windows. The fine church of Brailes is partially in this style, and even in some of the village churches, as at Wootton Wawen and Aston Cantlow, good specimens may be found, but perhaps the finest of all is the chancel (all that was built) of the Collegiate Church of Astley. The great church at Solihull has also Decorated work of high merit, including a groined crypt of unusual character.

The Black Death stopped church building till near the end of the century. Then a sudden outburst of fresh energy resulted in the creation of a new style, peculiar to England, that called Perpendicular. In this style the mouldings are flat, the arches flattened, the window tracery and wall panels largely in rectangular lines. It is usual to hear churches in this style denounced, but there is much to commend it, and the finest churches of Warwickshire are noble specimens. Foremost stand the two superb churches of St Michael and Holy Trinity in Coventry—the former with one long unbroken vista from the west to the east ; a nave and chancel enlarged by a succession of chantry chapels. Its finest feature is, however, its magnificent panelled tower and spire, a landmark on a clear day for many miles round. The churches of Aston-by-Birmingham and of Knowle are also fair examples of the period. The chancel of Stratford church is a late example of the style, but the one gem, unmatched either in the county or outside it, is the beautiful chapel built in St Mary's,

Holy Trinity Church, Coventry

Warwick, by Richard de Beauchamp, Earl of Warwick, to receive his remains. His magnificent tomb with its effigy of gilded brass lies in a building created at enormous cost, which still retains many of its beautiful statuettes and some glorious glass. The delicate lace-like carving of

The Beauchamp Chapel, St Mary's, Warwick

the small side chapel, especially in its niches and fan-tracery roofs, can hardly be matched.

After the Reformation church building ceased, and the architecture of the time must be sought in domestic buildings. On the return of Charles II a new revival of church life began; a natural reaction against the

Puritans led to the repair or rebuilding of some churches during the seventeenth century, among the earliest of these being the little church of Compton Wynyates, a curious structure with a double aisle. More important are the tower and nave of St Mary's, Warwick, reconstructed after the fire of 1682, and the late renaissance churches of Billesley and Honington. During the Civil War the domestic chapel at Arbury was consecrated by Archbishop Sheldon, and decorated with a marvellous ceiling by Martin of London and carvings attributed to Grinling Gibbons. The churches of the eighteenth century—the fine church of St Philip's, Birmingham, erected in 1711, alone excepted—are not worthy of much attention.

16.   Architecture—(*b*) Military.

The county is rich in remains of castles and of fortified manor houses, some of them of the highest importance. Warwick, Kenilworth, and even Tamworth are known by name at least to most persons who have any knowledge of English history; Warwick is still a dwelling place; Kenilworth, unfortunately, a ruin; and Tamworth a town museum.

Castles fall naturally into two distinct types—the Norman and the Edwardian. In the Norman the principal strength lay in a keep, arranged either as a strong and solitary tower, or as a wall about the margin of a raised mound. Such a keep had one or more courtyards or base courts attached to it, themselves fortified by mounds and walls. The gate-house was also strongly defended, and moats, sometimes dry, ran round the whole line of works. The plan of the Edwardian castle was introduced from the East, and it is termed concentric. It consisted of one or more complete lines of defence, courtyards with walls flanked at intervals by towers, each tower capable of resistance if the court and walls about it were taken by assault. Kenilworth and Warwick combine the two systems, the earlier castle having been added to from time to time.

When the invention of gunpowder rendered the castle proper no longer safe, castellated manor houses were built, and were strong enough to resist a short siege, thus giving the besieged time to admit of help arriving. Castles of this type still remain at Astley and at Maxstoke.

There are no castles in the county with existing remains dating back to the time of William I; indeed it was not until the reign of Henry II that stone-built castles became common. Before that date the earlier palisaded mound and base court (motte and bailey) had merely been strengthened, but of these earlier earthworks there is no lack in Warwickshire. The Montforts' castle at Beaudesert was of this type, and does not appear ever to have had much stone-work. The Cantelupes had a small castle at Aston Cantlow, and both Kenilworth and Warwick were at first defended by stockades alone. The moated mounds at Morton Bagot, Overberrow, and Ratley are further examples.

Warwick Castle owed its origin to the campaign

against the Danes. In the early summer of 913 Aethelfleda and her Mercians built Tamworth, and in the
following year Warwick, both having a mound and base
court. On the same site early in the reign of Henry II
the first stone castle of Warwick seems to have been
built. This had a shell keep of polygonal form, as at
York; but the walls now crowning the mound are of
much later date. Soon after the third quarter of the
fourteenth century Guy de Beauchamp, Earl of Warwick,
occupied his retirement by commencing the magnificent
courtyard as it now stands. The work was carried on
by his successor on the south-west side crowning the rocky
bank of the river, where there is a line of vaulted under-
crofts, with a hall, solar, and buttery above them. This is
guarded by the elegant trefoil-shaped "Caesar's Tower"
built, it is said, between 1350–1376, and forming a
remarkably strong but graceful defence of the bridge.
At the opposite angle "Guy's Tower," named after the
mythical hero "Guy of Warwick," was built in 1394.
It is twelve-sided, and consists of a series of vaulted rooms
having smaller apartments on either side, the loopholes
flanking the north-easterly and north-western ramparts.
Between these towers, protected by the barbican, stands
the gate-house, with its double doors, and double port-
cullis. A dry moat ran round the walls, and was crossed
by a drawbridge. The north-western ramparts were the
weakest portion, and here two towers, flanking a postern
gate, were commenced but probably never quite finished.
Yet another tower covered a postern gate on the south-
west side and defended the ramparts leading from the

Guy's Tower, Warwick Castle

mound to the main building.   In time of siege temporary
scaffolding was fixed to the inner side of the ramparts
and the corbels still remain.   This scaffold also overhung
the wall, and was doubtless defended by sheepskins, etc.,
from fire.   The castle narrowly escaped destruction in
the reign of James I, but was fortunately sold to Fulke

Kenilworth Castle

Greville, who restored it in excellent taste.   During the
Civil War it was more than once besieged and underwent
considerable damage.

Kenilworth was a stronger, larger, and even more
historically important castle than Warwick, and like
that great fortress was built on an early site admirably
suited by natural position for defence.   Here Geoffrey

de Clinton, second Earl of Huntingdon, built about
1180 a late Norman keep in the angle of the inner
ward. The walls of this keep are still 87 feet high and
from 13 to 14 feet thick at their base. Attached to the
keep is a "fore building" covering its approach. The
keep itself had corner turrets of some size, but these with
its north wall are almost gone, the latter having been
destroyed during the Civil War. The inner bailey is
chiefly occupied by what are called "Lancaster's buildings," which lie along the north, west, and much of the
south sides, and were erected principally by John of
Gaunt late in the fourteenth century. They consisted of
the kitchens, the "Strong Tower," and the magnificent
hall, the noblest in England, 90 feet long by 50 wide,
and supported upon a vaulted cellarage. The hall was
approached by a broad stairway.

The flimsy but lofty ruins of "Leicester's buildings"
succeed, and were erected, together with a great gate-
house, after Queen Elizabeth had granted the castle to
Robert Dudley, Earl of Leicester in 1570. The inner
bailey commanded and overlooked the works of the outer
bailey, which covered about nine acres. They were
carried out principally by Henry III, who also constructed
the dam which formed the upper lake. Kenilworth in
its glory, defended by its strong double line of defensive
walls and its wide lake and marsh, must have been almost
impregnable.

Maxstoke Castle is of later date. It was built by
William de Clinton in 1346, and is rectangular in
plan, guarded by a tower at each corner. It is entirely

surrounded by a moat, and access is obtained by a drawbridge leading to an elaborate gate-house which still
retains its original gates.

Astley Castle is a strongly fortified manor house,
surrounded by a deep moat, with a courtyard in front
defended by a broad wall, and entered by a drawbridge
leading to a gatehouse.

Astley Castle, Nuneaton

The other Warwickshire castles have little or no
remains of stonework.   Fulbrook Castle was a brick-
built edifice on the crest of the hill.   A portion of stone
walling of early date remains near it, and may have been
part of an earlier work.   Milcote Castle, the old seat of
the Grevilles, was burnt almost immediately after its
rebuilding during the Civil Wars, and about the same
time Wormleighton, the fortified home of the Spencer
family, was also badly damaged.

Little Wolford Hall

17. Architecture—(c) Domestic.

The houses and cottages of the county are almost entirely built of local materials, of stone roughly dressed from the nearest quarry, and either thatched or roofed

Leicester Hospital, Warwick

with thin stone slates. Such cottages, farms, and small manor houses are found wherever the material is available. In the better-wooded districts the houses are usually timbered or half-timbered ; in the latter case the lower storey is of stone. Instances of this may be seen at Little Wolford Hall, the ancient home of the Ingram family, and a house adjoining the churchyard at Hampton-in-Arden. Both examples are picturesque, and date from the reign of Henry VI. Of timbered houses many fine examples are still standing in the streets of Coventry, though some of the best are destroyed. There are a few in Birmingham, others at Knowle, Itchington, Stratford, and Warwick. The beauty of outline of the high-pitched gables, combined with the harmonious colouring of the stone roofs is very pleasing. Happily two remarkable examples still stand—the Leicester Hospital at Warwick, consisting originally of the joint buildings and halls of the wealthy guilds of St George and St Mary ; and the guild buildings at Stratford-on-Avon, now the grammar school. Both these examples are fairly complete, and retain their halls.

Throughout the country districts the humbler tenements show, in many cases, their ancient character, with low side walls of wattle and daub and high-pitched roofs covered with massive thatch, but they are fast disappearing. Of farmhouses but few really striking examples remain. There are examples at Hampton-in-Arden, and at Knowle ; and more pretentious stone houses in the Edgehill district, where such villages as Wormington and Ratley and Tysoe are almost entirely stone built. Three Warwickshire manor houses excel in the beauty of their brickwork. The finest is that at Compton Wynyates, the seat of the Earl of Northampton, which was built in the reign of Henry VI. It stands about a courtyard within an enclosure formerly moated and has beautiful chimneys of finely-moulded brick. Almost of the same date and equally interesting is Baddesley Clinton Hall, the home of the Ferrers family ; and still better known, Charlecote, the seat of the Lucy family, whose connection with Shakespeare, real or otherwise, brings numerous sightseers past its gates.

Warwickshire possesses many fine houses of a later

Compton Winyates

date, when the Gothic spirit in architecture had become dead. In the reigns of Elizabeth and James I a revival of classical taste, first noticeable as early as the reign of Henry VIII, began, and very rapidly houses arose in all directions. At Barton-on-the-Heath the Underhill family built a stately mansion from designs by Inigo Jones; at Billesley Hall another of considerable size and excellent panelling was erected by Sir Robert Lee, a London merchant; and the fine manor-house at Aston near Birmingham was planned by Sir Thomas Holte about the same period. This is now the property of the Corporation of Birmingham, and is perhaps as interesting as any in the county, since it is not only a fine architectural structure, but bears traces of the sharp attack it under-

Bilton Hall

went in the Civil War. Bilton Hall near Rugby, built in 1623, is also of historical interest from its connection with Addison, who bought it in 1711. Arbury Hall, an Elizabethan house, was largely altered by Sir Roger Newdigate, one of the pioneers of the Gothic revival; and, of later date, the nineteenth century Gothic house at Weston, the seat of the Earl of Camperdown, is worthy of note.

18. Communications—Roads, Railways, Canals.

The primitive savage required ready means of communication with the rest of his kin, and found it in devious forest tracks which gradually became more marked until well-defined paths appeared. These early pathways lie at the basis of our modern network of roads.

The oldest roads in the county are no doubt those along the crests of its hills. The best known of these is still called the Ridgeway, and from it an extensive view can be obtained. It forms part of the county boundary and leads from Evesham to Redditch; and almost parallel, but on the other side of the Arrow, runs another ancient road called the Portway. Both of these may have been altered by the Romans. Yet another Ridgeway runs along the Edge hills, linking Nadbury Camp with the earth-fort of Brailes and the Rollright Stones.

Another part of the pre-Roman road system is found in the Salt-ways. Both the Upper and Lower Salt-way pass through part of the county, leading from the great salt-producing centre of Wich or Droitwich. The Upper Salt-way apparently passed a little south-east of Birmingham on its way to Lincolnshire; the Lower ran from Droitwich through Feckenham Forest on its way to Cirencester, crossing the Ridgeway to Hanging Well near Coughton and over Alcester Heath to join the Icknield Street, which it follows through Bidford and Bickmarsh into Gloucestershire. This road must have had several branches, one running through Coughton and Great Alne to Stratford and on towards London, another deviating from it through Bearley to Warwick.

There is also an ancient way known as Pig Lane passing along the summit of the Ilmington Hills from the

The Icknield Way, near Wixford

direction of Bidford and ultimately joining the Fosse-way. Some, if not all of these roads were altered and adapted by the Romans, their paths leading to the best fords.

Three Roman roads cross the outer edges of the county. There is the North and South road through Alcester, or rather the series of roads leading from the

Fosse-way at Bourton-in-the-Water to Derby. The Warwickshire part can be easily traced, either as a main road or field tracks, except through the city of Birmingham, where there is no longer any sure clue to its direction. Part of its course forms, or nearly forms, the present county boundary. It ran from Studley to Alcester and from that town to Bidford, where it crossed the Avon by a ford near the church. There is no space here to discuss the names given to various parts of this road or their cause, but south of Alcester it still retains the ancient name of Buckle Street[1], which may have been that by which it was known to the Teuton conquerors. A branch road ran from Alcester to Stratford, where it forded the Avon and possibly continued along its left bank up the Stour valley, while another branch ran towards Banbury, both following the lines of older trackways. Another road ran from Alcester to Droitwich along the line of a Salt-way.

The Watling Street, which forms so large a part of the north-east boundary of the county, was so called in Saxon times. It ran from London past St Albans to Wroxeter. It enters the county a little to the south-east of Rugby and ultimately fords the Tame at Fazeley to enter Staffordshire. At Atherstone large irregular stones were found imbedded in it, some of which are said to bear marks of chariot wheels, but they do not seem to be suitable paving stones even if the road was ever paved, which is extremely doubtful.

The Fosse is the name given in Saxon times to a

[1] Bucgan or Buggilde Street.

series of roads which ran from Bath to Lincoln, and like the Watling Street its Warwickshire course is quite plain, much of it being still in use as a main or secondary road, and the remainder preserved as a field track. It enters the county on the north-east at High Cross, passing by Brinklow and Chesterton, where earthworks occur, probably in the former case of a later date. It crosses the Stour at Halford, and finally leaves the county at Stretton-on-the-Fosse. A short distance further, at Dorn, is the site of what may have been a Roman villa, with a drinking booth by the way-side.

One might expect to find a series of Saxon roads connecting such strongly fortified positions as Studley, Morton Bagot, Oldberrow, Beaudesert, and Fillongley with Tamworth, Warwick, and Kenilworth, but some of these important earth-forts have been entirely overlooked by local antiquarians, and the road system of that date has not been studied. The roads in the medieval period were little cared for ; indeed, there was practically no carriage traffic, such trade as there was being carried on by means of pedlars and pack-horses. The earlier roads were deep in dust in summer and mud in winter. Wains bearing corn to the religious houses must have been severely tried ere they reached their destination.

From the departure of the Romans until the year 1285 the roads of the county were repaired by the land-owners through whose lands they passed. This system sufficed when cartage was rarely resorted to. The main roads of medieval date led from one town to another and radiated from Coventry and Warwick.

A general Act for the improvement of the highways of England was passed in 1555, but the most conspicuous change was effected by the erection of tollgates or turnpikes in 1663. They were probably not introduced in Warwickshire until a century had passed. Thus the road from Alcester to Stratford was turnpiked in 1753, and the roads out of Birmingham altered and much diverted in 1793-1809. Turnpikes were abolished by the Act of 1872. John McAdam's system was introduced after 1819, and has continued in use, the material now chiefly employed by the county authorities coming largely from the diorite quarries in the Clee Hills.

In close connection with the roads of the county stand the bridges. These until the fourteenth century were probably constructed of wood and were only suitable for pack-horses. The old 14-arch wooden bridge at Warwick gave way to the stone building that is now a picturesque ruin. Some time later the monks of Bordesley built Bidford bridge, which is the best specimen now remaining. Halford bridge across the Stour has been much altered, as has that across the Avon at Binton, but the finest of all is the great bridge of 14 arches built across the Avon at Stratford by Sir Hugh Clopton in the reign of Henry VIII.

Canals existed in England in the reign of Henry VIII. At Stratford-on-Avon Andrew Yarranton spent much money in making the river Avon navigable, but the necessities of commerce, owing to the prosperity of Birmingham, brought the canal system of the county into being. The waterway from Birmingham to Coventry

The Old Bridge, Stratford-on-Avon

was cut in 1790, from Birmingham to Worcester in 1791, from Birmingham to Warwick in 1793. Communication was further aided by the invention of tramroads; and such a road, designed as part of the "Central Junction Railway" was laid down from Stratford-on-Avon to Moreton-in-the-Marsh in 1820 and opened in 1826.

Victoria Square, Birmingham

The invention of steam traction for railways in 1814 led to the gradual neglect of the canals, and this unfortunately still continues, the traffic upon them having fallen off to a very large extent. Birmingham is naturally the modern railway centre, with two principal large stations, the Great Western Station at Snow Hill, and the London and North Western, but Coventry is nearly as important.

Lines were laid from Birmingham to Derby in 1839, to Gloucester in 1840, to Rugby and Leamington in 1851, to Wolverhampton in 1852. Coventry was united to Leamington by rail in 1844, but the Grand Junction line from Birmingham to Newton opened in 1837, and is the oldest line in the county, though in the same year the great Midland town was joined to Liverpool, and in the following year to London.

19.   Administration and Divisions.

The scattered homesteads of the Teuton settlers had in early days little need of courts for the adjustment of their affairs.   At first each home was self-governing, the earliest approach to common action was the division of a district into "hundreds" and the meeting of a special court, usually supposed to have been a folk assembly of 100 free families.   The meeting-place was either a Moot mound as at Knightlow and Stoneleigh, or an enclosed space as at Pathlow, or even some conspicuous tree.   From the Hundred Courts representatives were sent to the Shire-Moot of the whole shire, which met but twice a year. In Domesday Book we find ten hundreds existing, but these were afterwards reduced to four.   The modern hundred of Hemlingford represents the Domesday hundred of Coleshelle.   Knightlow includes the hundreds of Bomelaw, Meretone, and Stanlei; Kineton those of Tremelaw, Honesberie, Fexhole, and Berriestone; and Barlichway the old hundred of Fernecombe.   In addition to these, there was the Bishop's Liberty of Pathlow, which survived well into the fourteenth century, if not later. Besides the Shire-Moot and Hundred Courts, the more important towns had courts of their own.   The burgesses had a right to assemble in a Town-Moot and elect Provosts or Bailiffs. Thus Alcester still has a High and Low Bailiff, as Stratford at one time had. These early officers gradually gave way to the more modern government by a Mayor, Aldermen, and Corporation. The corporate towns of the county are Birmingham, Coventry, Leamington, Nuneaton, Rugby, Stratford, Sutton Coldfield, Tamworth, and Warwick.

The city of Coventry was incorporated in 1345 and made a separate county in 1432, but the Charter was cancelled in 1842. Birmingham was raised to the rank of a city in 1838, and its presiding magistrate is now styled "Lord Mayor."

The various manors all over the county had also courts of their own.   In theory these courts were either the Court Baron, or the View of Frankpledge. The former dealt with grants of land to those who held land "by Copy of Court Roll," and represented the private jurisdiction of the lord of the manor ; the latter represented the King's interest in local affairs and was in theory the Court of the Sheriff.

Warwickshire was represented by two members of Parliament for the county and two each for the boroughs of Coventry and Warwick, making a total of six.   Birmingham had no representative until the passing of the Reform Act in 1832, when it had two seats assigned to it, but these were raised to three in 1867 and there are now seven for the city and one for Aston. Warwick and Coventry had also two each and there were four County Members to represent the shire, namely one each for the Northern or Tamworth Division, the North-eastern or Nuneaton Division, the South-western or Stratford-on-Avon Division, and the South-eastern or Rugby Division.

The principal officer of the county is the Lord Lieutenant. Since 1757 that office has been held with the office of Custos Rotulorum; under him are the Sheriff (the ancient Shire-reeve) and his deputies. The administration of the county was in part carried out by the Quarter Sessions, its last chairman, John Stratford Dugdale, appointed in 1883, becoming first Chairman of the County Council. The County Council has control of finance, roads and bridges, the county buildings, allotments, sanitation, police, education, diseases of animals, etc. It takes the place of the ancient Shire-Moot and is composed of 17 Aldermen and 53 Councillors. It governs the whole of the administrative county, which coincides with the geographical county with the exceptions of parts of the County Borough of Birmingham, the County Borough of Coventry, the Borough of Tamworth, and the Local Board Districts of Hinckley and Redditch lying within the county.

Local government of towns and parishes was formerly carried on by vestries, highway boards, local boards, and so forth. These, by the Act of 1894, were replaced by District Councils, Parish Councils, and Parish Meetings. For the relief of the needy there are 17 Poor Law

Unions in the county, each with its Board of Guardians, who manage the workhouses and all that pertains to the relief of the poor and the aged who are not eligible for the Old Age Pension scheme.

The Quarter Sessions for the county are held at Warwick and Coventry by adjournment. Besides these there are 14 Petty-Sessional Divisions, each provided with magistrates or justices of the peace, whose duty it is to try petty offences. Birmingham has courts of its own.

Warwickshire lies in three dioceses, and here, as elsewhere, the Church was organised and its system established before that of the State. The whole county was originally in the great Mercian diocese, with Lichfield for its cathedral town, itself a member of the Archiepiscopal See of Canterbury. St Chad was the first Bishop of Mercia who chose Lichfield as his see, A.D. 664, but Duima, A.D. 656, was first Bishop of Mercia. In 673 the Council of Hertford ordered this great diocese to be divided, chiefly at Archbishop Theodore's wish. The south part of Warwickshire peopled by the Hwiccas was joined with their Worcestershire kinsmen, and Bosel, a monk of Whitby, became first bishop in 680, with Worcester as his cathedral town. The dual diocese then settled down to the gradual formation of the parochial system by which the bishop has a priest in every village acting for him, and responsible for the spiritual well-being of the people and the proper conduct of divine service. North Warwickshire soon after the Conquest had a cathedral of its own at Coventry, Robert de Limesey removing his see to that town in 1192. Only four bishops styled themselves

Bishops of Coventry, their title after 1183 running "of Coventry and Lichfield." The cathedral of Coventry was ultimately destroyed after the Dissolution, to the great loss of the county. Lastly in the year 1905 Charles Gore, then Bishop of Worcester, became first Bishop of Birmingham; that see being carved out of Worcester and Lichfield.

The bishop has certain officials called archdeacons whose duty chiefly concerns finance. Of these there are three in the county, one for Warwick, one for Coventry, and a third for Birmingham. The clergy are also grouped in what are called rural deaneries, the rural dean being usually chosen by the bishop, who appoints an incumbent in the deanery. It is his duty to report to the bishop, and to convey from the bishop matters he desires should reach his clergy. As in the case of the archdeacon it is also his duty to visit parish churches and report upon their condition.

20. Roll of Honour.

The list of Warwickshire worthies is a long one, and includes the greatest dramatic poet of the world, William Shakespeare, the splendour of whose genius has thrown into the shade the light of lesser men.

The only cathedral city within the county in early days was Coventry, the seat of a bishopric which for most of its existence has been united with the mother see of Lichfield. Among the great bishops of this joint see we find Walter de Langton, who died in 1321, one of the many distinguished men who rose from humble origin to become powerful. Keeper of the Great Seal of England, he was also a trusted councillor of Edward I, and at one time Lord Treasurer of England. At a later date a famous Protestant divine, George Abbot (1562–1633), one of the translators of the Authorised Version of the Bible, was for a short time bishop, and was translated to the Archiepiscopal See of Canterbury in 1610. Another bishop, John Overall (1560–1619), was one of his colleagues in translating the Bible, and held the post of Regius Professor of Divinity in Cambridge. He was chosen Bishop of Coventry in 1614, and translated to Norwich four years later.

Several natives of the county rose at various dates to episcopal rank. Two of the most important were Robert de Stratford, Bishop of Chichester, 1337, and John de Stratford, Bishop of Winchester, 1323–1333, afterwards translated to Canterbury, a prelate who was largely responsible for the government of the country during the absence in France of King Edward III. Among other cele-brated bishops born in Warwickshire were John James, Bishop of Calcutta, born at Rugby in 1786; John Bird Sumner, Archbishop of Canterbury, born at Kenilworth in 1780; Richard Parre, Bishop of Sodor and Man, born in 1592; and John Ryder, Archbishop of Tuam, born at Nuneaton, 1697. Of the clergy who did not reach episcopal rank the most important were John Kettlewell (1653–1695), vicar of Coleshill, widely known as "Coles-hill's Saint," who was ejected from his benefice as a non-juror, and wrote a number of devotional works;

Thomas Wagstaff, born at Binley near Coventry, who composed *A Defence of King Charles the First*; and John Trapp (1601–1669), master of Stratford School and vicar of Weston-upon-Avon, who wrote a *Clavis to the Bible*, and was much esteemed as a scholar in his day.

Birmingham, being exempt from the five-mile Act, since it was not legally speaking a town, was, as we have already seen, a stronghold of Puritanism. Its most noted nonconformist in the eighteenth century was Dr Priestley, of whom we shall speak later, but there have been many noted teachers elsewhere in the county, among them Richard Baxter (1615–1691), connected with Alcester and Coventry. He took a prominent part in the Savoy Conference, and one of his many devotional works, *The Saints' Everlasting Rest*, is still read. Another famous Puritan divine, Thomas Cartwright (1535–1603), was Master of Leicester Hospital in Warwick, and author of many works. The number of "ejected ministers" turned out of their benefices at the Restoration was great.

A few natives of Warwickshire have risen to high posts in the legal profession. Among these may be named Sir Orlando Bridgeman (1606?–1674) of Coventry, a royalist lawyer who held the post of Chief Baron of the Court of Exchequer and presided at the trial of the Regicides. Sir John Puckering, M.P. (1544–1596), was Chief Justice of South Wales in 1585, and Speaker of the House of Commons. He was also for a time Lord Keeper of the Great Seal, and died in 1585. Sir Edward Saunders of Weston-under-Wetherley died in 1576, one of the illustrious men whom Queen Elizabeth attached

to her service. He rose to be Lord Chief Justice. Sir
John Willes, D.C.L., a native of Bishop's Itchington,
where he was born in 1685, became Lord Chief Justice
of the Court of Common Pleas. Another Lord Chief
Justice was Sir John Eardley-Wilmot, 1709–1792, who,
though born at Derby, was a member of a well-known
Warwickshire family.

Warwickshire has also had its share of distinguished
medical men, among them some of early date. John
Hall of Stratford-on-Avon (1575–1635), best known as
the husband of Susanna, daughter of William Shakespeare,
was an educated and well-travelled man, the author of
Select Observations on English Bodies. He had no medical
degree, but obtained considerable reputation as a prac-
titioner in his town. James Cook of Warwick was the
author of *The Marrow of Chirurgery*, and Thomas
Dover of Barton-on-the-Heath (1662–1742), the dis-
coverer of the uses of ipecacuanha and of "Dover's
Powder," practised chiefly in London. William Withering
(1741–1799), physician to the General Hospital in Birmingham, was widely distinguished as a botanist and
mineralogist. Lastly Edward Johnstone, M.D. (1757–
1851), one of the first physicians of Birmingham
General Hospital, and President of Queen's College,
was one of the founders of the Queen's Hospital in
that city. John, his brother (1768–1836), another of
the distinguished medical men of Birmingham, and a
writer on medical matters, was known as an intimate
friend of Dr Parr of Hatton. Both brothers were born
at Kidderminster.

The high position of Rugby School in the educational world brought men of great note into the county to occupy the position of Headmaster, the foremost of whom was Thomas Arnold, D.D. (1795–1842), the reformer of

Thomas Arnold
(after the portrait by Thomas Phillips)

public schools in England. He introduced a system by which the boys were trusted largely with their own control. This raised in them a sense of honour, which has ever since been a marked feature of public school

Interior of Shakespeare's House

training. Among other illustrious masters of Rugby School were Thomas James, D.D. (1748–1804), and Dr Temple, afterwards Archbishop of Canterbury.

The county has produced but few statesmen of note, if we except Sir Nicholas Throckmorton (1513–1570), who deserves mention as one of Queen Elizabeth's ambassadors.

During the Civil War a number of commanders arose. On the Parliamentarian side, Colonel William Purefoy of Caldecote (?1580–1659), M.P. for Coventry, won infamy as a great defacer of churches, including the Beauchamp Chapel at Warwick. He was also one of the Regicides. One unfortunate navigator, Sir Hugh Willoughby, who died in 1554, should not be forgotten. He set out with a small fleet to discover a north-east passage to Cathay, and perished miserably of starvation and scurvy with his men on the coast of Lapland.

The poets of the county are headed by the greatest poet of the English race, William Shakespeare, born at Stratford-on-Avon on April 23rd (O.S.) 1564, son of John Shakespeare, a burgess of Stratford of good repute in the town, and Mary Arden, daughter of a yeoman farmer of Wilmcote. The birth is supposed to have taken place in the house in Henley Street, now known as "the Birthplace." The future poet probably attended the grammar school, then held in the Chapel of the Guild, under a master named Walter Roche. In 1582 he married a certain Anne Hathaway, who lived at Shottery close by; but the marriage was apparently not a very happy one. According to tradition Shakespeare became

involved in a poaching affair at Charlecote, and to avoid
prosecution left the town about 1586. He went to

Anne Hathaway's Cottage

London and seems to have obtained some small employment at one of the new theatres. He then became a
member of the Earl of Leicester's company of actors,

8—2

and acted with them from 1592-9, taking parts in his own plays, and also in those of other authors. In 1594 he performed before the Queen. During the 12 years of his London life he lodged in Silver Street, with a certain Frenchman named Christopher Mountjoy, who was a fashionable wig-maker. In 1597 he returned to his native town, and bought New Place, the largest house in it, perhaps from the profits of his share in the Globe Theatre, of which he was a joint lessee. He then returned to London, having rescued his family from want, and did not permanently return to Stratford until 1611. He was frequently in London until 1614, and kept up his association with the stage to the end. Shakespeare died at New Place on the 23rd April (O.S.), 1616, and was buried before the altar of the parish church, where a bust was erected to his memory before 1623. His first plays were those which he altered or amended for use in the company of players of which he himself was a member. It was only later that the great tragedies of *Hamlet*, *Macbeth*, and *Othello* came from his pen.

A contemporary and friend of Shakespeare ranks next him in the roll of fame, Michael Drayton, a native of Hartshill (1563-1631). Like the greater poet he wrote historical pieces, but is best known for a long poem descriptive of the counties of England, called the *Polyolbion*. Drayton rests in Westminster Abbey. Another writer of this date is Sir Fulke Greville, Lord Brooke (1554-1628), a statesman, courtier, and friend of Sir Philip Sidney, whose life he wrote, together with a number of poems and a tragedy called *Mustapha*.

Bust of Shakespeare, Holy Trinity Church,
Stratford-on-Avon

William Somervile (1677–1742), a poet of a later school, was a member of the Edstone family of that name. He was a sportsman and country gentleman who devoted his spare moments to literature, and is best known for his poem *The Chase*. Walter Savage Landor (1775–1864), was a native of Warwick, an enthusiast of great power combined with a quarrelsome nature. He lived in many places, Bath, Bristol, and Wells, and later in life in Jersey, France, and Italy. A number of works, both in prose and verse, came from his pen, but the ablest are the series of papers styled *Imaginary Conversations*. Nicholas Brady, D.D. (1650–1726), vicar of Stratford-on-Avon, was the joint author with Nahum Tate of a metrical version of the Psalms, formerly used in most churches, which appeared in 1696.

The historians of the county include the honoured name of the antiquary, Sir William Dugdale (1605–1686), who was born at Shustoke. Rouge Croix Poursuivant, and later Chester Herald, he was a stout Royalist, and in 1642 was called upon to attend the Earl of Northampton and array in arms the men of his county. In the course of his herald's office he summoned the castles of Banbury and Warwick to surrender, and performed good services to the Royal cause. He was with the King's army in Oxford and during his stay in that city he made great collections towards his antiquities of the county, and these were published in 1656. He lived in retirement during the Commonwealth, but after the Restoration he continued his valuable work, and published the Baronage of England. He was then created Garter Principal

King of Arms in 1677, and died in 1685. His history of Warwickshire is the principal authority upon all that relates to the past of its great families and estates, and his famous *Monasticon* gives the history of the English religious houses.

William Hutton (1723–1815), the historian of Birmingham, was a friend of Dr Priestley, Baskerville, and the rest of the Birmingham circle of distinguished men of that period. Like his friend he had his house burnt in the riots of 1791. Thomas Sharpe, the historian of Coventry (1771–1841), wrote an excellent history of his town, and Robert Bell Wheler (1785–1857), a lawyer and antiquary, performed the same good work for Stratford.

Finally, Warwickshire can claim one of the greatest of English novelists in Marian Evans (1819–1880), who wrote under the name of George Eliot. This most brilliant of writers was born at Arbury Farm in Chilvers Coton, where her father was estate steward. Her novels for the most part depict, with a fidelity which is unsurpassed, the life of the middle classes in the midlands, and *Adam Bede*, *The Mill on the Floss*, and *Silas Marner* reach the highest level of excellence, as indeed, though in a different field, do *Middlemarch* and *Romola*.

Neither Priestley nor Baskerville were Warwickshire by birth, but both these great men were largely connected with the county. Joseph Priestley (1738–1804), whose greatest discovery was that of oxygen gas, came to Birmingham in 1780, and was one of the most distinguished chemists of his time ; and John Baskerville (1716–1775), beginning life as a footman, became in turn schoolmaster, stone-mason, a painter of japanned goods, typefounder, and lastly the most famous Birmingham printer, who

George Eliot
(*from the portrait by Sir F. W. Burton*)

turned out editions renowned throughout Europe for their excellence.

Matthew Boulton (1728–1809), engineer and inventor,

the joint founder with James Watt of the famous
Soho works, was yet another of Priestley's friends.

David Cox

(*from the portrait by William Radclyffe*)

Boulton had much to do with the improvement of the
coinage, and invented the milling machine.   He was also

responsible for the copper coinages of 1797 and 1809. James Watt, LL.D., F.R.S. (1736–1819), a native of Scotland, began life as an organ-builder, but practically invented the steam engine in 1774. He rose after a hard struggle, overcoming enormous difficulties, but from the date of his partnership with Boulton his success was assured.

Turning to music and art we find William Croft (1677–1727), who was born at Nether Ettington, the most distinguished name as a musician. He began life as a chorister of the Chapel Royal, of which he ultimately became organist, and afterwards organist of Westminster Abbey. His anthems are much admired. Birmingham near the middle of the last century became a well-known school of water-colour painting. Here, at Deritend, was born David Cox (1793–1859), who ranks as one of the greatest of British water-colourists, and Birmingham was also the birthplace of Sir Edward Burne-Jones (1833–1898) pre-Raphaelite, friend of William Morris, equally master of oil and water-colour, as well as designer of stained glass and tapestry.

21. THE CHIEF TOWNS AND VILLAGES OF WARWICKSHIRE.

(The figures in brackets after each name give the population in 1911, and those at the end of each section are references to the pages in the text.)

Alcester (2168), a small market town on the river Alne, eight miles westward of Stratford-on-Avon. It possesses a town hall built in 1641, and a church containing a fine tomb in memory of Sir Fulke Greville and his wife, Lady Elizabeth Willoughby. There was formerly an Abbey here, but of this there are no remains. (pp. 20, 26, 49, 58, 71, 77, 99, 101, 105, 110.)

Alveston (905), a parish adjoining Stratford-on-Avon, of which it is rapidly becoming a suburb. The disused church has two remarkable Norman doorheads. The monks of Worcester had a grange here at Bridge End, which still retains some good timber work. (pp. 71, 75.)

Ansley (127), a village five miles north-west of Nuneaton. Coal and ironstone are worked in the parish. The old hall is still standing. (p. 51.)

Aston Cantlow (896), an agricultural parish with several hamlets, 5½ miles north-west of Stratford-on-Avon, noticeable for the excellent window tracery of the Guild chapel in the church, and earthworks of a "castle" of the Cantelupe family, as also for many timbered cottages. (pp. 82, 86.)

Aston-juxta-Birmingham, now incorporated in the city of Birmingham, to the north of which it lies. The church contains a series of tombs of the early owners. Aston Hall, a magnificent structure built by the Holte family, is used as a museum. (pp. 82, 96.)

Atherstone (5607), a market and manufacturing town, five miles north-west of Nuneaton. The church is of small interest, but there is an ancient grammar school, and manufactories of hats. The Watling Street passes through the town. (pp. 6, 46, 61, 77.)

Balsall or **Temple Balsall** (1353), two miles east of Knowle, noteworthy for its church, built entirely in the Early Decorated style and rising by steps gradually from east to west, and for the almshouses founded by Lady Katherine Leveson in 1670. (p. 77.)

Barcheston, a small village one mile south-east of Shipston-on-Stour. Here was founded by a country squire, William Sheldon, in the reign of Henry VIII, one of the most remarkable tapestry works in England.

Barford (694), a village three miles south of Warwick. The bridge over the Avon and the ancient tower of the church are its most interesting features. The latter is said to have been a mark for the Puritan cannon as the troops passed on their way to Edgehill. (p. 38.)

Bedworth (9595), a large mining village three miles south of Nuneaton. Besides its coal and ironstone mines, there are factories of hats, tape, ribbons, etc., brick kilns and iron-works, and important canals. (pp. 25, 51.)

Berkeswell (1577), a beautiful village six miles west of Coventry. The church possesses a curious crypt and also a timbered vestry over the porch, there is also a hall in the parish, once the seat of the Wilmot family, whose monuments are in the church.

Bidford (1634), a village with several hamlets, formerly a market town, four miles south of Alcester. There is an ancient bridge here across the Avon, built by the monks of Bordesley, and some interesting stone houses, among which is the Falcon Inn, associated by tradition with Shakespeare. (pp. 7, 18, 97, 98, 99.)

Bilton (5188), a village adjoining Rugby. The hall gates bear the initials of Joseph Addison and his wife the Countess of Warwick, who lived here. The more populous part of the parish New Bilton, has important brick and cement works. (p. 71.)

Birmingham (526,000). This great city extends its borders into Worcestershire and Staffordshire and in a sense belongs equally to all. The nucleus is however in our county. It is a city of peculiar origin as it rose out of a large village and until the nineteenth century had no Corporation. It was then an open town. It is to-day an admirably administered city and possesses a flourishing University, Municipal Art galleries, and a large public library. The old parish church in the Bull Ring has been admirably restored and is a beautiful building containing an interesting series of tombs of the De Berminghams, who built a castle here in 1154. This family obtained grants of a market and fairs for their village out of which small beginning its present prosperity has arisen. This is mainly owing to its proximity to the rich mineral deposits of Warwickshire and Staffordshire aided by its healthy position and important canal trade, and later by its still more important railway connections.

St Philip's Church, used as the Cathedral of the Diocese, is a fine specimen of the architecture of the early 18th century; it contains some beautiful painted windows designed by Burne-Jones.

Among many important buildings is the Town Hall, built in the Corinthian order from designs by Harris. It is in this splendid hall that great political gatherings are usually assembled, and near it are erected statues of Queen Victoria, King Edward VI and various local celebrities. The Corporation buildings, of remarkably handsome character are close at hand as is the Art Gallery and Public Library. (pp. 15, 16, 38, 49, 54–60, 68, 78, 85, 94, 96, 97, 99, 101, 103–108, 110, 111, 119, 120, 122.)

Bishop's Itchington (818), a village with some fine timbered houses three miles south-west of Southam, formerly a manor of the bishops of Lichfield, but now chiefly noted for its cement works. There are the remains here of an episcopal manor house. (pp. 20, 111.)

Brailes (809), a large parish three miles east of Shipston-upon-Stour. The church has a lofty tower in which is a well-known ring of bells, the tenor weighing 35 cwt. There is also a seventeenth century grammar school, and an early Roman Catholic mission, which possesses some valuable ancient vestments and a library of sixteenth century books. There are earthworks in the parish of peculiar form; one, a long bank, runs due east and west at the summit of the hill, another a circular work is known as "Castle Hill." (pp. 71, 81, 97.)

Brinklow (667), a large village 5½ miles north-west of Rugby. The church, in the Early English style, rises 12 feet on the floor level. There is a curious mound here of early but uncertain date. (p. 100.)

Budbrooke (1104), a village 1½ miles west of Warwick, contains the barracks and headquarters of the Warwickshire regiment. Grove Park, the ancient seat of Lord Dormer, is also in this parish.

Bulkington (1837), a large village six miles north-east of Coventry, contains a fifteenth century church which belonged to the Abbey of Leicester. Within this are a number of carvings from the chisel of Richard Hayward, a local sculptor, who gave them to the church about 1790. Ribbon weaving is the staple industry.

Burton Dassett (475), a large village and formerly a market town, four miles east of Kineton. The church is built on the hill-side, and the floor rises by steps eastward. It is mainly in the Norman and Early English styles, and retains some particularly graceful capitals. There are ironstone quarries in the parish, and an ancient beacon on the crest of the hill. (pp. 14, 15, 30, 52, 76.)

Chesterton (130), an agricultural village 6½ miles south-east of Leamington, memorable for its "Roman" camp on the Fosse-way, and a windmill of peculiar form, said to have been designed by Inigo Jones, who built the house of the Peyto family here. Of this latter only a brick gateway remains. (pp. 20, 33, 100.)

Chilvers Coton (10,492), a large and rapidly-increasing parish adjoining Nuneaton. Temple House, a grange of the Knights Templars, and Arbury Hall are within its bounds. The latter has extremely beautiful grounds, and a range of stabling erected from designs of Sir Christopher Wren. In the hamlet of Griff was born Marian Evans, who wrote under the name of "George Eliot." There are extensive coal mines in the parish, and large drain-pipe and tile works. (pp. 50, 53, 119.)

Claverdon (520), a large agricultural and residential village six miles west of Warwick. There is a stone tower here, an adjunct of a mansion commenced by Thomas Spencer, but never finished. (p. 32.)

Clifton-upon-Dunsmore (627), an extensive parish two miles north-east of Rugby. The church, built in the thirteenth century, had at one time a "fair steeple," but it was taken down to save expense. In the parish is Cave's Inn, a Roman site. (p. 18.)

Coleshill (2886), a small market town eight miles east of Birmingham. The pillory, whipping post, and stocks are standing. The church is a fine one, mainly Decorated in style, with lofty spire and rich in monuments of the Clintons and Digbys. The font is Norman. At Marston Green, in the parish, is a home for children of the Birmingham poor. (pp. 8, 10, 16, 104, 109.)

Combe Fields (156), a liberty and parish five miles east of Coventry, containing Combe Abbey, the beautiful seat of the Earl of Craven. (pp. 20, 77.)

Coughton (206), a large village two miles north of Alcester. There are some curious earthworks here called "Danes Banks," and a stately mansion erected by the Throckmorton family in the reign of Henry VIII. The church contains the tombs of this family and some interesting stained glass. (pp. 97, 98.)

Coventry (106,349[1]), an ancient city and formerly a cathedral

Ford's Almshouses, Coventry

town, arising about a Priory founded by Earl Leofric in 1043. It is now a busy manufacturing centre, turning out large supplies of bicycles and motor cars. Ribbon-weaving and watch-making were formerly its staple trades. Its wealthy merchants built in the Middle Ages the glorious church of St Michael, one of the

[1] The population at the present time is believed to be about 180,000.

largest parish churches in England, and that of the Holy Trinity
its companion. Coventry was constituted a county by itself in
1451, but this was abolished in 1843. It is, nevertheless, a
Parliamentary borough, returning one member. It was also a
cathedral city in place of Lichfield from 1075 to 1188, in which
year the style of its bishops was changed to "Lichfield and
Coventry." It will probably again become the seat of a bishopric
at an early date. The Corporation possess St Mary's Hall, which
contains some valuable tapestry and stained glass. Ford's Almshouses are fine examples of half-timber architecture. (pp. 6, 15,
25, 38, 44, 49, 50, 54, 56–61, 63–70, 77, 78, 82, 94, 100, 105,
106, 108, 110.)

Curdworth (335), a parish three miles north-west of Coleshill. The church dates from the Norman period, and contains
the remains of a pre-Norman font.

Dunchurch (935), a large parish, a town three miles south-
west of Rugby. The church is principally in the Decorated style.
The ancient hundred court of Knightlow is still held here and its
curious custom of wroth-silver preserved.

Ettington or **Eatington** (570), a scattered agricultural
parish 5½ miles south-east of Stratford. Its most noteworthy feature
is the beautiful park of the Shirley family, famed for its ancient
hawthorns, and the ruined church containing the tombs of the
family and some old glass said to have come from Winchester
College. In the park are two early camps; one of these is locally
called Roman, the other is far earlier. (pp. 10, 20, 26, 27, 53.)

Exhall (1646), a village 3½ miles north of Coventry. There is a
large and important colliery here, and also ironstone works. (p. 51.)

Fenny Compton (510), a village 11 miles south-east of
Warwick with a station on two lines of railway. The church dates
from the fourteenth century, and there are many stone-built
houses, one of which, with an early fourteenth century window of
much merit, is said to have been a leper hospital.

Fillongley (1425), a very large village six miles north-west of Coventry. The church is a large one with a spire, chiefly in the Decorated style, but containing Norman remains. There are several mansions in the parish. (p. 100.)

Foleshill (7781), a parish adjoining Coventry, formerly a Chapelry of the Priory there. The church is a fifteenth century building. There are manufactures of ribbons, fringes, and elastic for hats.

Hampton-in-Arden (1084), a large village with many hamlets, three miles east of Solihull. The church, which belonged to the Priory of Kenilworth, dates from Norman times, and has a spire. The "Castle Hills" are the remains of one of the many fortified seats of the Arden family. (pp. 15, 16, 37, 38, 94.)

Harbury (1160), a large village five miles south-east of Leamington. Its modern prosperity is due to its lime and cement works, which are very extensive. The church belonged to the Priory of Kenilworth, but has no special features of interest. (pp. 26, 53.)

Hartshill (2450), a mining village three miles north-west of Nuneaton, chiefly to be remembered as the birthplace of Michael Drayton the poet. There is an important colliery here, and the ruins of a small castle. (pp. 25, 71.)

Hatton (1524), a village three miles north-west of Warwick, the scene of the labours of the famous Dr Samuel Parr. The County Lunatic Asylum is situated on its border, and the inmates are included in the number of the population given above.

Henley-in-Arden (1062), a market town eight miles north of Stratford-on-Avon, and formerly a Chapelry of Wootton Wawen. The Guild House and the remains of the market cross are the most noteworthy objects, but there are many timbered houses of interest. The chapel is of late Perpendicular work and of no great merit. (pp. 8, 58.)

Hillmorton (1259), a scattered village 2½ miles south-east of Rugby. The church contains some interesting tombs, one of them attributed to Sir Thomas de Astley, who was killed in the battle of Evesham. There is a stepped cross in the village. (p. 8.)

Ilmington (576), a large village in the hollow of the Cotswolds, four miles north-east of Chipping Campden. The church dates from the twelfth century with a rude Norman tower and arches. There is a stone manor house recently restored. The oolite is quarried to a small extent in the village for building purposes. A pretty valley called Lark Stoke, about a mile to the west, contains the remains of the mansion of the Brent family, and a medicinal well, at one time of considerable repute. (pp. 10, 28, 30, 32, 44, 53, 70.)

Ipsley (907), a large parish to the east of Redditch, part of which town is within its limits. The church has some early incised slabs of the Huband family. The inhabitants are chiefly engaged in the manufacture of needles.

Kenilworth (5776), a town five miles north of Warwick, famous for the ruins of its castle, and for its wealthy priory, a few ruined walls of which remain near the church, which dates from the Norman period, and possesses a rich twelfth century western doorway. It is rapidly increasing in size, serving for many Coventry workers, who find it difficult to obtain houses in the city. (pp. 5, 25, 63–68, 76, 77, 78, 85, 86, 90, 100, 109.)

Kineton (1018), a small town which formerly had a market, nine miles south-south-east of Warwick. The church dates from the Early English period, and has a good western doorway enriched with dog-tooth moulding. The kennels of the South Warwickshire hounds are situated in the hamlet of Little Kineton. There is an earthwork here called "King John's Castle." The battle of Edgehill was fought close by. (p. 104.)

Kingsbury (3831), an extensive parish on the Tame five miles north-east of Coleshill. There is some late Norman work in the church. The Hall here was at one time the seat of the Bracebridge family. There was also an ancient beacon on the hill.

Knowle (2357), a township and parish 9½ miles south-east of Birmingham, noticeable for its Guild buildings and collegiate church of fifteenth century date. (pp. 78, 82, 94.)

Lapworth (853), eight miles west-north-west of Warwick.

The Old Well House, Leamington

The church is a good example of fourteenth century work, but the Norman clerestory can be traced. Robert Catesby the conspirator was born in this parish.

Leamington (26,713). An important town and spa, two miles east of Warwick, on the Leam, which owes its prosperity to its medicinal waters. It has a pump room, assembly rooms, beautiful ornamental gardens, various institutions, and a large residential population. The parish church is a good example of modern work, built on the site of the original church of the village. The borough includes the parishes of Lillington and Milverton. (pp. 18, 104, 105.)

Leek Wootton (481), a village 2½ miles north of Warwick, of much natural beauty. In this place is the mansion called Woodcote, and at Chesford an ancient bridge over the Avon.

Long Compton (547), a large village six miles south-south-east of Shipston-on-Stour, is chiefly noticeable as containing Weston Park, the seat of the Earl of Camperdown, built on the site of the ancient home of the Sheldon family. Just beyond the parish boundary stands the fine stone circle called the Rollright Stones. The "King's stone" is within the Warwickshire boundary. (p. 8.)

Long Itchington (1178), a large village five miles east of Leamington. The church dates from the fourteenth century, and belonged to the monks of Hertford. (pp. 33, 94.)

Mancetter (703), a parish one mile south-east of Ather-stone, of which it is the parent, is chiefly of interest as a Roman site. The manor house is of early date and of unusual con-struction. (p. 9.)

Maxstoke (230), on the Blythe, three miles south-east of Coleshill, possesses important remains of a priory, including a gate-house with its original gate, and a castle built by William de Clinton in the reign of Edward III. (pp. 77, 86, 90.)

Meriden (832), 5½ miles north-west of Coventry, has a church standing on the crest of the hill, noticeable chiefly for its monuments. The "Woodmen of Arden," a society of archers, meet at Forest Hall in this village.

Monks Kirby (544), an important parish six miles north-west of Rugby. The church, which contains monuments of the Fielding family, had a very lofty spire, which was taken down to avoid expense in 1630. (pp. 15, 19.)

Napton-on-the-Hill (847), three miles east of Southam. The church occupies a commanding position on the summit of the hill, and contains much good work of the twelfth and thirteenth centuries.

Nether Whitacre (606), three miles north-east of Coleshill. In the church, which dates from the thirteenth century, is

Rugby School

a monument to Charles Jennens, a native and a benefactor who endowed a free school in this place. (p. 16.)

Newton Regis (450), a village five miles north-east of Tamworth. The church has a lofty spire, and contains some good Decorated window tracery. (p. 8.)

Nuneaton (37,073), an important town nine miles north of Coventry containing a number of factories in which cotton goods, elastics, hats, and woollen goods are made, with collieries in the hamlets. The remains of the Abbey of St Mary have been incorporated in a modern church. The town has grown very rapidly, and was incorporated in 1909. (pp. 6, 9, 15, 24, 25, 50, 53, 61, 77, 91, 105, 109.)

Olton, a district of Solihull, 4½ miles south-west of Birmingham, chiefly famous for its beautiful and extensive reservoir, which is visited by many rare birds. (pp. 20, 32.)

Polesworth (5619), a small town four miles north-west of Atherstone. The church formed part of the ancient Abbey, and contains two early effigies of its abbesses. The parish is a colliery one, and there are also clay works. The market is no longer held. Pooley Hall, built by Sir Thomas Cockayne in 1509, retains much of its ancient walling.

Priors Marston (495), a large village five miles south-east of Southam, has an Early English chancel with well-proportioned windows to its church. It belonged anciently to the Priory of Coventry.

Rowington (868), a large village six miles north-west of Warwick. The church is of fifteenth century date with a square central tower, and there are some ancient half-timbered houses. (p. 52.)

Rugby (21,758), a market town, formerly a Chapelry in the mother parish of Clifton-upon-Dunsmore. The school was founded in 1567 by Laurence Sheriff, but was raised to its high position by a series of eminent masters, among them the famous Dr Arnold and Dr Temple. It is becoming an important manufacturing centre. (pp. 6, 18, 26, 38, 49, 53, 61, 96, 99, 104, 105, 112.)

Salford Priors (823), a large village four miles south of Alcester at the confluence of the Arrow and the Avon. It possesses an interesting church, a marked feature being the turret stair on the south and some curious monuments in memory of the Parker family.

Shirley, a large village, formerly a member of Solihull, five miles south-east of Birmingham. The church was built in 1532.

Snitterfield (682), a parish three miles north-north-east of Stratford. It possesses an early fourteenth century church, which contains some bench-ends elaborately carved with the arms of Henry VIII. (p. 12.)

The Grammar School, Stratford-on-Avon

Solihull (10,282), a town seven miles south-east of Birmingham, contains one of the finest churches in the county, dating from the Early Decorated period, the work being curious and unusual. The chapel of St Alphege has a crypt below it, and is of great beauty; its original altar remains. It has many side chapels and a tower with a modern spire 213 feet in height. The village has several imposing mansions, and has become a residential suburb of Birmingham. (p. 82.)

Southam (1804), a small market town six miles south-east of Leamington. The church has a tower of thirteenth century date, but is chiefly in the Decorated and Perpendicular styles. It was from this town Charles I set out for Edgehill. (p. 53.)

Stockingford (9708), a hamlet 1½ miles west of Nuneaton formed into an ecclesiastical parish in 1843. There are several large collieries and extensive manufactures of tiles, blue bricks, and drain-pipes. (p. 25.)

Stockton (975), a village two miles north-east of Southam with large cement works. The chancel and tower of the church are of Decorated date.

Stoke (51), a suburb of Coventry, on the eastern side of the city, with a church dating from the thirteenth century. In this place in 1471 was born William Hollis, Lord Mayor of London. (p. 53.)

Stoneleigh (1400), a large village on the Avon four miles south of Coventry. Stoneleigh Abbey, the seat of Lord Leigh, retains considerable portions of the ancient abbey. The parish church dates from the Norman period and has a curious early font. There is a moot hill opposite it, where a Court of Pleas was formerly held. (p. 104.)

Stratford-on-Avon (8531), a large town, chiefly of interest as the birthplace of William Shakespeare. The birth-place, and the site of New Place, in which he died, are carefully tended and used as museums, etc. The stately parish church in which the poet was buried dates from the thirteenth century, work of this date being used in the reconstruction of the tower, and is charmingly situated. Among other objects of interest in the town are Clopton Bridge of 14 arches over the Avon, and the ancient buildings of the Holy Cross Guild, a fine range of black and white construction used as a grammar school, probably the only buildings of that date built for school purposes and still used.

The house called Harvards and another known as Tudor House have carved beams and brackets. (pp. 3, 10, 18, 19, 38, 54, 59, 70, 72, 78, 79, 80, 82, 94, 98, 99, 101, 103, 105, 116.)

Studley (3019), a village four miles north of Alcester, anciently contained a small priory, of which there are some remains. It is now a centre of the manufacture of needles, fish-

The West Gate, Warwick

hooks, and fishing tackle. Studley Castle, on the hill near, is a modern building, converted into an Agricultural College for women. The moat of the old castle may be seen near the church. (pp. 20, 58, 77, 99.)

Sutton Coldfield (20,132), an ancient market-town seven miles north-east of Birmingham; the birthplace of John Voysey, Bishop of Exeter, who obtained a grant of the park here for the corporation of the town. This park contains 2400 acres, and is a favourite locality for botanists and entomologists. The parish church dates from the thirteenth century, but has many later alterations, including an eighteenth century nave. (pp. 31, 32, 105.)

Tanworth (2231). An important parish four miles north-west of Henley-in-Arden. The church dates from the fourteenth century. The parish contains large canal reservoirs, and a populous hamlet called Salter Street. (pp. 19, 38.)

Warwick School

Tysoe (751), (Upper, Middle, and Lower), a group of three villages some 4½ miles south of Kineton. The church is of early date, some portions were probably built in the eleventh century, including the windows above the nave arcade. The village has many stone houses of unusually picturesque character. A gigantic figure called the "Red Horse," which gave its name to the valley, was cut on the hill-side opposite the church, and was supposed to commemorate the battle of Towton, the Kingmaker's victory. It has unfortunately entirely disappeared. (pp. 76, 94.)

Warwick (11,858), the county capital and centre of its administration. The Shire Hall and jail are here, and a museum rich in local geological specimens. Two of the town gates remain and part of the wall. Both of these gates are surmounted by

Parcloses, Wootton Wawen Church

chapels, that on the west gate being used by the pensioners of the adjoining Leicester Hospital, an institution which took over the halls and buildings of the earlier guilds. The magnificent castle of the Earls of Warwick is placed on the south of the town, dominating the approach from Banbury. St Mary's Church contains the burial place of the Earls of Warwick; on the south is a chapel called the Beauchamp Chapel, a gem of elaborate Perpendicular work, built to receive the tomb and effigy of gilded bronze of Richard de Beauchamp. It has also monuments to Robert Dudley, Earl of Leicester, and Ambrose, Earl of Warwick, and considerable remains of valuable stained glass. The old leper chapel of St Michael, now a blacksmith's shop, has a good fourteenth century window, and the mansion erected on the site of the Priory of St John of Jerusalem is noticeable for its wrought-iron gates. Owing to a great fire in 1694 the town is chiefly modern. Warwick is also famous for its School, which dates back to the days of the Saxon Collegiate Church of All Saints, which stood within the Castle precincts. It is without doubt one of the oldest of our English schools. (pp. 3, 5, 18, 25, 30, 31, 32, 39, 54, 57, 59, 63–68, 73, 77, 78, 85, 86, 87, 89, 93, 94, 98, 100, 101, 105, 106, 108, 110.)

Wellesbourne (1358), a double village, distinguished as Wellesbourne Hastings and Wellesbourne Mountford, five miles east of Stratford-on-Avon. The church, greatly injured in restoration, has an Early English circular window in its south aisle, and a brass of Sir Thomas le Strange in full plate armour. (p. 70.)

Wolvey (657), a village five miles south-east of Nuneaton. The church dates from the Norman period, and contains some interesting monuments to the Astley and Wolvey families. There is a tradition that Edward IV was surprised on Wolvey Heath by the Kingmaker. Wolvey Hall was erected about 1676 on the site of an earlier edifice.

Wootton Wawen (1959), a large and scattered village six miles north-west of Stratford-on-Avon. Its church is a particularly interesting building, the tower is pre-Norman with " long and short work," and other portions of this date occur in the nave walling. It contains the most perfect chancel screen-work in the county, the parcloses being well preserved. The side chapel has a window decorated with good ball-flower work. (pp. 20, 75, 82.)

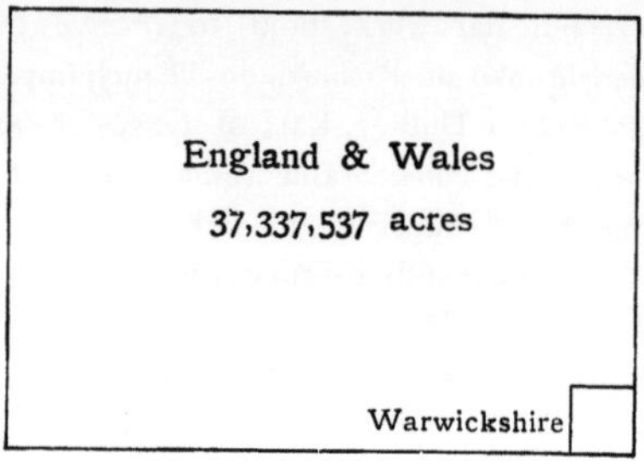

England & Wales

37,337,537 acres

Warwickshire

Fig. 1.　Area of Warwickshire (605,273 acres) compared
with that of England and Wales

England & Wales

36,070,492

Warwickshire

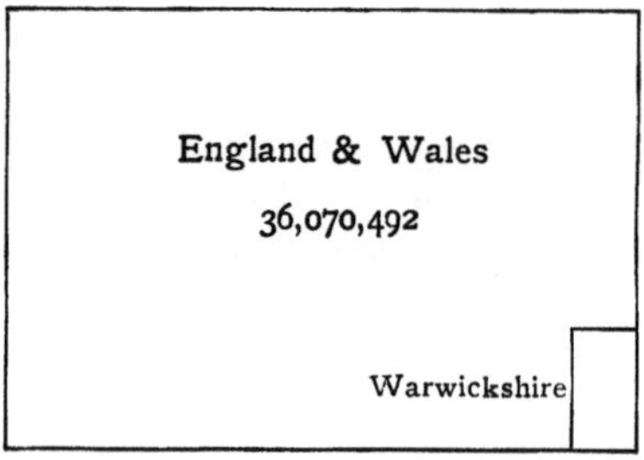

Fig. 2.　Population of Warwickshire (1,040,409) compared
with that of England and Wales in 1911

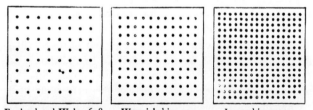

England and Wales 618　　Warwickshire 1100　　Lancashire 2554

Fig. 3.　Comparative density of population to the square
mile in 1911

(Each dot represents 10 persons)

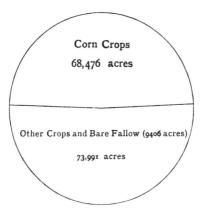

Fig. 4. Proportionate Area under Corn Crops compared with
that of other cultivated land in Warwickshire in 1913

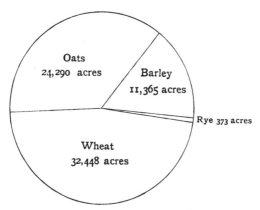

Fig. 5. Proportionate Areas of chief Cereals
in Warwickshire in 1913

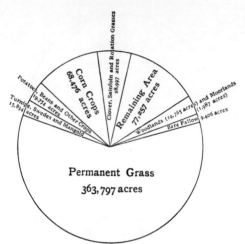

Fig. 6. Proportionate Areas of land in
Warwickshire in 1913

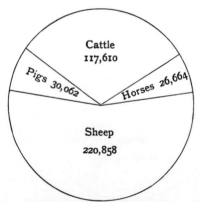

Fig. 7. Proportionate numbers of Live Stock
in Warwickshire in 1913